Adorning Bodies

Also available from Bloomsbury

Adornment: What Self-Decoration Tells Us About Who We Are,
by Stephen Davies
Beauty and the End of Art, by Sonia Sedivy
The Aesthetic Illusion in Literature and the Arts, edited by Tomáš Koblížek
The Changing Boundaries and Nature of the Modern Art World,
by Richard Kalina
The Visibility of the Image, by Lambert Wiesing

Adorning Bodies

Meaning, Evolution, and Beauty in Humans and Animals

Marilynn Johnson

BLOOMSBURY ACADEMIC
LONDON • NEW YORK • OXFORD • NEW DELHI • SYDNEY

BLOOMSBURY ACADEMIC
Bloomsbury Publishing Plc
50 Bedford Square, London, WC1B 3DP, UK
1385 Broadway, New York, NY 10018, USA
29 Earlsfort Terrace, Dublin 2, Ireland

BLOOMSBURY, BLOOMSBURY ACADEMIC and the Diana logo are trademarks
of Bloomsbury Publishing Plc

First published in Great Britain 2022
This paperback edition published 2023

Copyright © Marilynn Johnson, 2022

Marilynn Johnson has asserted her right under the Copyright, Designs and Patents Act,
1988, to be identified as Author of this work.

For legal purposes the Acknowledgements on pp. ix-x constitute an extension of this
copyright page.

Cover image: John Gould and H.C. Richter (British, 1804–1881), Docimastes
Ensiferus (Sword-billed Hummingbird), Photo by: Universal History Archive/
Universal Images Group via Getty Images

All rights reserved. No part of this publication may be reproduced or transmitted in
any form or by any means, electronic or mechanical, including photocopying, recording,
or any information storage or retrieval system, without prior permission in writing from
the publishers.

Bloomsbury Publishing Plc does not have any control over, or responsibility for, any
third-party websites referred to or in this book. All internet addresses given in this
book were correct at the time of going to press. The author and publisher regret any
inconvenience caused if addresses have changed or sites have ceased to exist, but
can accept no responsibility for any such changes.

A catalogue record for this book is available from the British Library.

A catalog record for this book is available from the Library of Congress.

ISBN: HB: 978-1-3501-0425-9
PB: 978-1-3503-0130-6
ePDF: 978-1-3501-0426-6
eBook: 978-1-3501-0427-3

Typeset by Deanta Global Publishing Services, Chennai, India

To find out more about our authors and books visit www.bloomsbury.com and sign
up for our newsletters.

Dedicated to my father, who taught me how to see nature

Contents

List of Illustrations		viii
Acknowledgements		ix
1	Meaning in Bodies and Adornment	1
2	Taking Adornment Seriously: Structuralism and Meaning	15
3	Details on the Gricean View	37
4	Deception in the Human and Animal Worlds: Imitation of Natural Meaning and Lying in Non-Natural Meaning	57
5	Darwin on Animal Bodies	71
6	Human Sexual Selection	87
7	The Evolution of Bodily Adornment: Signaling and Meaning-Making in Prehistory	111
8	Emotions: Information, Misperception, Suppression, and Expression	139
9	On Beauty: Aesthetic Choices, Adornment, and Art	159
Notes		183
References		189
Index		204

Illustrations

Figures

1	Women's hat designed by Elsa Schiaparelli, 1949	17
2	Lorena, Paul, and Mary Beth Tinker, of Tinker v. Des Moines, 1969	21
3	Ugly sweaters in Amsterdam	41
4	Items seized from Jeremy Wilson's apartment	58
5	Hailee Steinfeld in Prabal Gurung at the Met Costume Gala, 2014	61
6	RuPaul at the MTV Video Music Awards, 1993	62
7	*Reclining Dionysos* from the Parthenon East Pediment, Elgin Marbles	64
8	Image of Lady Diana Spencer on the steps of Saint Paul's Cathedral, 1981	82
9	*Portrait of a lady*, by Rogier van der Wayden, *c*.1460	112
10	Statue of King Akhenaten at Karnak, *c*.1365 BC	114
11	Statuette of King Akhenaten and Queen Nefertiti at Tell el-Amarna, 1353–1337 BC	116
12	Sample stimuli: 13-point morph continua, with varied race and status attire	155
13	A satin bowerbird at Lamington National Park in Queensland, Australia	163
14	A display of Schiaparelli fashions from an exhibition at the Victoria and Albert Museum in London, 1971	178
15	Rihanna at the Met Gala for Rei Kawakubo/Comme des Garcons: Art of the In-Between, 2017	180

Tables

1	Summary of Types of Meaning	68

Acknowledgements

This book came to fruition with the encouragement of many supportive mentors and friends. Because it is my first book, my journey to completing this book included my path into philosophy more broadly. I would like to thank some of my first professors Harry Brighouse, Alan Sidelle, and Brynn Welch at the University of Wisconsin-Madison for showing me what philosophy could be and for their encouragement about going to grad school.

I wrote the first drafts of these chapters while completing my PhD at the City University of New York Graduate Center. I would like to thank a number of people for their support during those years in New York, including Daniel Bender, Noël Carroll, Chris Caruso, Michael Devitt, Suki Finn, Peter Godfrey-Smith, Jenny Judge, Kathleen Hefty, Louise Lennihan, Stephen Neale, Nick Pappas, Chelsea Peng, Graham Priest, Jesse Prinz, Ashley Ryan, David Rosenthal, Alexandra Schaumber, Cameron Tung and Iakovos Vasiliou. Part of this book was written with funding provided through the American Society for Aesthetics Doctoral Dissertation Fellowship. They ask that I note that any views, findings, conclusions, or recommendations expressed in this publication do not necessarily reflect those of the American Society for Aesthetics.

I am also grateful for the friendship and guidance of many people in Miami, where I had a postdoc and was offered the contract for this book: Sean Allen-Hermanson, Otávio Bueno, Elizabeth Cantalamessa, Allan Casebier, Clinton Castro, Molly Castro, Aja Edwards, Caleb Everett, Simon Evnine, Pamela Geller, Michael Grafals, Anton Killin, Marygrace O'Hearn, Hermes A. Perdomo, Elizabeth Scarbrough, Laurie Shrage, and Paul Warren.

The ideas presented in this book were sharpened in response to audience questions in presentations given at a number of institutions: the Interuniversity Center in Dubrovnik Croatia, the Institut für Philosophie, Freie Universität Berlin, Germany, the University of Queensland, Brisbane, Australia, the University of Sydney, Australia, LIM College, New York, the Cognitive Science Seminar at the CUNY Graduate Center, New York, Florida International University, Miami, Florida, the Florida Philosophical Association Meeting at the College of Central Florida, Ocala, Florida, the Society for California Archaeology at Joshua Tree National Park, California, the Philosophy

Colloquium at the University of Nevada-Reno, and the American Philosophical Association Eastern Division Meeting in Savannah, Georgia. At times in a footnote in the book I will note how my thinking about a question at such a presentation or in the conversations that followed made its way into the book. I would especially like to thank Francesco d'Errico, Michael Glanzberg, Simone Gubler, Dunja Jutronić, Elmar Unnsteinsson, Mary Stiner, Una Stoinić, and Ben Young, for their incisive questions or for helping me arrange these presentations.

The final stages of this book were completed while at my current position as Assistant Professor at the University of San Diego. I would like to thank the university for Faculty Research Grant support to dedicate time to my writing. I would also like to thank colleagues and friends in San Diego for their enthusiastic support of the project while completing the book: Corey Barnes, Mark Chapman, Brian Clack, Ryan Doody, Kristin Moran, Noelle Norton, Luis Osuna, Eliza Smith, Hermes Taylor, and Nick Riggle.

I would like to thank the excellent team at Bloomsbury for their work on the book. From the early stages of the contract, through the review process, to the design of the cover, and through copyediting they have been a model of professionalism and a joy to work with. I would like to especially thank Colleen Coalter, Louise Dugdale, Becky Holland, Suzie Nash, Benedict O'Hagan, and Mohammed Raffi. I would also thank two anonymous reviewers for Bloomsbury whose comments helped improve the final manuscript.

The book has also benefitted enormously from conversations and exchanges with scholars that were conducted online, and in a recently published symposium in *The Journal of Aesthetics and Art Criticism*. I would like to thank Wesley Cray, Stephen Davies, Thi Nguyen, Jonathan Neufeld, and Evelleen Richards, for their correspondence about the project. Lastly, I would especially like to thank Michelle Johnson, who read and copyedited the entire manuscript. Any errors that remain are mine.

1

Meaning in Bodies and Adornment

Introduction

We adorn our bodies. We wear ties, undergarments, wigs, toupees. We emphasize a jawline, hide a hairline. We modify our bodies. We make some parts of our bodies appear larger or smaller. We lift weights. We diet. We eat. We wear boots, bras, suits. Some parts of our bodies persist throughout our lifetimes. Some change. Some bodies give birth. Some bodies do not. Eventually our bodies grow old. Our bodies die.

Whatever the self is, it is bound up with the body—in this lifetime anyway. We all walk around with bodies, as bodies. We are perceived, judged, ignored, policed, respected, deferred to, or questioned because of these bodies we live in, live as. Our bodies have a height, a weight, a skin tone, are gendered, and placed into racial categories on the basis of how they are perceived. The ways our bodies are perceived depends on whose gaze we are under. The meanings attributed to certain bodies have changed throughout time and will continue to change.

Our bodies and the ways we adorn them are taken to *mean* things. In this book I consider this meaning. How do we do this? Why do we do this? Where did it come from? Is meaning in bodies like other types of meaning? Is it like clouds on the horizon that mean rain? Is it like language? Can we be wrong about what bodies mean? Should these meanings be changed? Can these meanings be changed?

Mind and Body

The seventeenth-century French philosopher René Descartes thought there was an "I," a consciousness, this "thinking thing" that *had* a body. The body served as a means by which we perceive the world, an idea some have later called "the ghost in the machine" (Descartes 1640/1984; Ryle 2000). The mind and the body were seen as distinct. We might agree with Descartes, and think the body is "a provisional residence of something superior—an immortal soul, the universal or thought" (Latour 2004). Or, we might think that Descartes was wrong—we do not *have* a body; we *are* a body.

Our consciousness is inextricably bound to our bodies for the time being, and to speak of the mind and body as distinct things might be seen as a metaphysical error. Commenting on Descartes philosopher Gilbert Ryle writes,

> The problem [of] how a person's mind and body influence one another is notoriously charged with theoretical difficulties. What the mind wills, the legs, arms and the tongue execute; what affects the ear and the eye has something to do with what the mind perceives; grimaces and smiles betray the mind's moods and bodily castigations lead, is it hoped, to moral improvement. (Ryle 2000: 12)

In a number of disciplines there have been recent shifts toward understanding the self *as* body, as this messy, warm, entity that gets dirty and bathes and jitters when it's nervous and develops ulcers under too much stress. This has impacted fields ranging from neuroscience to psychology to linguistics to archaeology (Van Der Kolk 2015; MacFarquhar 2018; Malafouris 2016; Johnson and Everett 2021). Today it is recognized that our strongest theories of emotion and consciousness are informed by theories of phenomenology and embodied cognition and place our experience as bodies at the forefront (Chalmers and Clark 1998; Prinz 2008; Johnson and Everett 2021).

While there has been this shift in philosophy of mind and other disciplines to considering the centrality of body, the branch of philosophy that engages most centrally with questions of meaning—philosophy of language—has not adequately engaged with consideration of bodies and bodily adornment. It is time that the questions about bodies and meaning I have posed here be considered with all the power and tools of analytic philosophy of language.

Interpretations

The ways our bodies appear are interpreted by those around us. Taller people are thought to be more natural leaders, for example (Judge and Cable 2004; Maclean 2019). People who are "babyfaced" are thought to be less competent (Zebrowitz and Montepare 2005). And, people judged to be more attractive are happier (Diener et al. 1995). Why is this? Where did these standards come from?

Bodies are also interpreted as signals in the non-human animal world. For example, male house finches have yellow, orange, or red heads and upper bodies, and the more varied and nutritious their diet, the deeper the tone (Tekiela 2000). House finch coloring comes from pigments, called carotenoids, found in the food they eat (Sundstrom 2017). Thus, their coloring serves as a reliable indicator of their ability to procure food. When bears are engaged in a fight, their hair stands on end, causing them to appear larger. Although our body hair has long since dwindled, we maintain this automatic epithelial response as well. Because in bears this creates a visual illusion about the size of the body, we might say that it is unreliable or a deceptive means of communication with the body.

The ways that we use our bodies to signal information have many parallels with animals. In this introductory chapter I will present a few case studies of communication by bodies, which I will return to later in the book. I will discuss reliable as well as unreliable bodily signals.

And although, as I'll later argue, some of what we use our bodies to mean is language-like, in other ways it falls outside of what we normally consider the realm of words or signs. Part of that gap is in the story of how adornment came to be. Our bodies evolved to be how they are because of the processes of natural and sexual selection, and this is a history we share with all other creatures. And so, our proper starting place in considering meaning in bodies and bodily adornment will not yet be with the philosophers of mind or language, but with Charles Darwin and the birds, a topic I will return to in the second half of the book.

A House Finch Communicates

While living in New York City in graduate school I bought a bird feeder. I lived in a second-floor apartment in Brooklyn and got a feeder that suction-cupped onto the glass. The window could open inward and I would shimmy my arms over the top to fill the feeder. A few minutes later it would be covered in birds. There were ledges on either side of the feeder and the mourning doves, due to their size, would get these prime spots in the face of any competition. Their wings would murmur a whir as they landed, and one on either side would gulp down sunflower seeds whole, as a third mourning dove waited in a queue on top for his turn.

Once the mourning doves finished, other, smaller birds came to my feeder. A common visitor was the red-breasted house finch. His color made him a pleasant surprise at the window. I soon noticed that there seemed to be a connection between the brighter-red-colored finches and their ability to secure a place on my feeder. Sometimes a duller red house finch would come along and I would see a brighter-red finch shoo him away, claim his spot on the feeder, and eagerly resume his feasting. New Yorkers are not known for their hospitality. I came to think of the brighter-red house finches as "meaner" and the duller ones as "weaker."

Upon consulting my *Birds of New York* field guide, I learned that house finches have a red color that is tied to their diets (Tekiela 2000). These house finches were redder because they had a more varied diet—a result perhaps of the same assertive behavior I was observing out of my window. I learned to see this red color as a sign that they would succeed in maintaining their place on the feeder ledge. In this way their coloration was a reliable indicator of their abilities (Sundstrom 2017). This is a signal that I could interpret and it could also convey the same information to males of the same species, who might choose to hang back and wait rather than needlessly expend energy in attempting to eat before a brighter-red male. The bright-red coloring of certain males also would signal this ability to procure a wide range of food to the female house finches that would choose a mate.

At the same time as I was making these observations out of my New York City window, I was completing my dissertation project on meaning, construed broadly. In this work I considered a range of different types of meaning, beginning with meaning in language, and from there considering meaning in literature, art, law, and archaeology. I argued there for a Gricean approach to meaning, as I will in this book as well. Although my focus at the time was primarily human communication, I couldn't help but see meaning in the house finches' coloring. If we understand meaning to be reliable communication of information about the world then the degree of redness certainly had meaning. But is this how we *should* understand meaning? Is all animal patterning like this? Does it always convey information? Is it always *reliable* information? And if it is communicative, who is the audience? How did it *become* meaningful? I stored these questions in the back of my mind for deeper analysis at a later time, the result of which is this book.

As I will discuss in detail in a later chapter, an explanation for the feathers of birds was at the forefront of Charles Darwin's mind throughout his career. Indeed, the very first chapter of *The Origin of Species* is devoted mostly to discussion of the feathers of domestic pigeons (Darwin 1859/2003: 20–40). Darwin's theory of natural selection explained how things in the natural world (and us as a part of it) evolved to suit the natural environments enough to procreate and pass on genetic material. We could understand the red pigmentation of house finches within this framework. As I will explain later, Darwin's contemporary Alfred Russel Wallace argued that when coloration is a reliable indicator of fitness it could be explained within the confines of natural selection. But, notably, not all animal coloration is an indicator of fitness.

Bodily Communication in *Homo sapiens*

To understand meaning in bodies and bodily adornment we must begin by understanding how the ways we use our bodies are analogous to non-human animals. We do, after all, share evolutionary history with non-human animals. Much of what we signal with our bodies is of this base, animalistic nature: we use bodies of others to gauge their sexual fitness, assess their physical strength, and to even judge them morally. Drawing on the work of philosopher of language H. P. Grice, this is what I will explain in terms of natural meaning. At the same time, we use and adorn our bodies in ways animals—even our primate cousins—do not. Certain forms of bodily adornment—such as an armband worn in protest of the Vietnam War—require recognizing the wearer's specific communicative intentions by mindreading. This is a cognitive capacity unique to humans. Again, drawing on the work of H. P. Grice, this is what I call non-natural meaning.

The ways we communicate with our bodies can sometimes be best understood within the broader context of how animals *in general—of which we are one—*communicate with bodies. At the same time, some communication with bodies—like some other communication—is unique to *Homo sapiens*. These are the sorts of communications with bodies that are most like our use of language or signs. I will argue for this thesis in the following chapters.

We must also keep in mind that the fact that a male house finch gets across information with his red coloring is completely outside his control. In this way his communication with his bodily coloring is different from how we communicate with words. Words evolve through cultural transmission, not genetic transmission.

When considering house finches, it is difficult to step outside of our own minds and our human preconceptions about behavior. We must be wary of attributing human mental states when interpreting animal behaviors. At the same time, drawing too stark a distinction between human and non-human animal behavior can cause us to lose sight of the fact that we, too, are animal. Some of what we do with our bodies is analogous to what animals do.

In developing his theory of sexual selection Charles Darwin argued along these lines, drawing on a metaphor between female mate choice in the animal world and women's choices in fashion. As I will detail in a later chapter, Darwin's analogy makes use of some dubious reasoning but its basic point remains: the same forces that have shaped the peacock's tail have also shaped us. Perhaps the way animals feel when they gaze upon each other is similar to how we feel (Prum 2013, 2017). It might seem like an indefensible thesis but as far as Darwin was concerned (Richards 2017), and as has been recently advocated by ornithologist Richard Prum (Prum 2013, 2017), this would mean that we and other animals share a similar aesthetic sense of beauty (Wilson 2016). I will consider this claim in the final chapter. Let me now commence with an overview of the chapters to follow, before making a few remarks on the aims of this book.

Overview of the Book

Chapter 2: Taking Adornment Seriously: Structuralism and Meaning

In the following chapter I will introduce us to bodily adornment and present previous attempts by philosophers to take it seriously. Consideration of adornment by philosophers is a relatively recent phenomenon. The ancient Greeks found adornment to be a paragon of frivolity. The most serious attempt to systematically study meaning in adornment was undertaken by Roland Barthes in the 1950s and 1960s. His project failed in ways he recognized, but considering his attempt and why it failed proves instructive. It also sheds light on broader problems with structuralism, the framework Barthes worked within.

Consideration of the reasons Barthes failed provides support for a more nuanced account of meaning. In presenting an account of meaning in bodies, as is the aim of this book, a distinction must be made between two different types of meaning. In this chapter I will then present and discuss natural and non-natural meaning. For this distinction I draw on the work of philosopher of language H. P. Grice. In his canonical 1957 paper "Meaning" Grice distinguishes between natural meaning and non-natural meaning. Natural meaning is meaning things in the world have, such as in "Those clouds mean rain" or "Those spots mean measles." Along these lines, we believe taller people are better leaders. We believe "babyfaced" people are incompetent.

In this sense we take bodies themselves to have meaning, which I will consider in terms of Gricean natural meaning. Natural meaning is also present in the animal world. I present a number of illustrative examples. Natural meaning is contrasted with what Grice calls non-natural meaning—meaning that we give and interpret on the basis of communicative intentions. I will present Grice's distinction. I then present a number of ways that we communicate with clothing that require us to consider intentions. As I argue, the way we communicate with bodily adornment parallels the way we communicate with words. If I wear a black armband, you know I am protesting an event. However, you must know something about my intentions to identify what it is I am protesting. We convey a number of complex messages with adornment by non-natural meaning. These are presented in Chapter 2.

Chapter 3: Details on the Gricean View

In Chapter 3 I will expand on and provide further details of the positive Gricean proposal that I have presented thus far. I will explain the distinction from philosophy of language between word meaning and speaker meaning and will discuss how this applies to adornment. I will present the case of the ironic sweater, which I understand to be an instance of Gricean implicature in dress. In this chapter I will also consider and respond to the following questions: Doesn't this mean anything can mean anything? Can't we be mistaken about meaning? What about personal stylists? What about uniforms? Are we "slaves" to fashion? Isn't this mere imitation? Isn't this really a matter of conventions? I will argue that all of these questions can be dealt with. I will also use Grice to assuage any potential worries about applying theory from philosophy of language to nonlinguistic communication. Finally, I will respond to those who insist that they say nothing with their clothing or don't think about what they wear. I will end the chapter with some words in favor of artifice as beauty, drawing on Baudelaire.

Chapter 4: Deception in the Human and Animal Worlds: Imitation of Natural Meaning and Lying with Non-Natural Meaning

With communication by our bodies and the ways we adorn our bodies comes the possibility of lying and deception with our bodies, the focus of Chapter 4. This applies to both natural and non-natural meaning. With non-natural meaning, the mechanisms by which we do this are very similar to ordinary language. If I am a con artist trying to get others believe I went to Yale Law, walking around with a Yale Law baseball hat will be a good start. If I would like to get others to believe I am a surgeon, I could buy some green scrubs and wear them to lunch. For this to succeed, I must hide the fact that I am attempting to deceive the viewer. Lying or deceiving with non-natural meaning parallels how we lie with language.

Lying or deceiving with natural meaning as I have noted, is a bit more complicated, and parallels again the animal world. For instance, when the fur of bears stands on end, this creates the visual illusion that they are larger and thus more physically imposing. I call such manipulation of bodies—and thus the associated meanings we have with bodies—imitation of natural meaning. Many of the ways we dress change the ways our bodies look and thus contribute to imitation of natural meaning. A case of just this is the suit, which has persisted for over 200 years because it appears to broaden men's shoulders, diminish their stomach, and lengthen their legs. This leads to a physical form that we have positive associations with. I present a number of instances of imitation of natural meaning in the chapter.

Chapter 5: Darwin on Animal Bodies

Next I turn to consider how meaning in bodies evolved. If we know we do these things with our body, how did we get here? Charles Darwin's theories relate to how certain physical traits are passed on from generation to generation. Some animal traits can be explained in terms of how they make a creature especially well suited to the natural environment around it. For instance, the long necks of giraffes can be explained in terms of the special advantage this provides for reaching the leaves of trees. In Chapter 5 I present Darwin's intellectual journey with the theory of natural selection and provide a number of concrete examples to illustrate the point along the way.

However, not all animal traits can be explained in terms of increasing suitability for the natural world. Darwin was famously irked by the peacock's tail. It clearly did not help the animal navigate its natural environment, and in fact was maladaptive, presenting a hindrance to behaviors such as procuring food, avoiding predators, and so on. To explain features such as the peacock's tail, Darwin developed his theory of sexual selection: the theory that certain features develop because they are attractive to potential mates. The selection pressures for sexual selection are often in tension with the selection pressures of natural selection. To this point, I also present Darwin's theory of sexual selection, exploring how it has been seen through the decades up to today.

Chapter 6: Human Sexual Selection

Understanding this dichotomy between the forces of natural and sexual selection is essential to understanding the ways we use our bodies to convey information. Discussion of sexual selection also brings up historic attempts to control and constrain bodies, especially those of women, people of color, and those who transgress the social norms of gender expectations. In this chapter I present examples of this, and discuss school dress codes, the Stonewall Riots, nightclub dress codes, and events called SlutWalks—which I take to be recent attempts at metalinguistic negotiation about the meaning of adornment.

Chapter 7: The Evolution of Bodily Adornment: Signaling and Meaning-Making in Prehistory

In this chapter I turn to consider bodily adornment as an evolved behavioral practice. Throughout the book I have been arguing for taking bodily adornment seriously as a conveyer of meaning. It is not something trivial and simple as it has been treated by many philosophers. This is never more apparent than when we consider how important bodily adornment is for the archaeological record, and for our understanding of who we are as a species. The ability to convey intentional, symbolic meaning has sometimes been presented as what distinguishes us from beast. Certainly this has been breathlessly argued with remains such as early cave art. But, in fact, the earliest proposed symbolic practices in the archaeological record are those of bodily adornment. In this chapter I present these accounts of our earliest symbolic behaviors found in the form of ochre that adorned the body and in shell beads. I argue that meaning in bodily adornment is a key piece into understanding who we are, where we came from, and our human practice of engaging in certain forms of symbolic behavior.

Chapter 8: Emotions: Information, Misperception, Suppression, Expression

In addition to our physical selves being taken to indicate persistent traits such as age or social standing, our bodies also reveal passing information about our emotions. As Darwin noted in *The Expression of the Emotions in Man and Animals* there is commonality in how we and our non-human relatives unconsciously signal our emotional states. Darwin observed how a dog that is happy will hold his ears in a different way when he is sad. He discusses our behavior of leaning in the direction we would like a pool ball to go. In certain parts of the world it is thought rude to honestly convey some emotions. For instance, in the West, one cannot show displeasure when opening a disappointing gift. Women walking in the street are sometimes told to smile, as though the fact that they may be concentrating is an affront to those they pass.

Certain emotional states can be very difficult to suppress. Because of this, they are unlike other ways we intentionally communicate with words. Further complicating the matter, the mechanisms underlying the relationship between our minds and our bodies is not as straightforward as we may think. It is not simply that our minds feel a certain way and this is revealed through the body. Our bodies send information to our brain as well, in a feedback loop. For instance, it has been shown that when participants in a study received undetectable subcutaneous Botox rather than a placebo between their eyes—where the brow furrows in concentration and anger—this resulted in them feeling happier. In this way, our bodies convey meaning not just to others but even to our own minds. In this chapter I consider these ways that bodies signal to ourselves, and close by considering how race affects perception of emotions and meaning of adornment in others.

Chapter 9: On Beauty: Aesthetic Choices, Adornment, and Art

In the final chapter, I will consider how the argument I have presented thus far relates to our broader notions of beauty and art. Ornithologist Richard Prum has presented an account of beauty in bodies and adornment in the human and animal world in which he argues that bodies constitute art. This is because he argues that the sexual selection mechanisms that lead to things like the peacock's tail constitute co-evolved aesthetic choices. Although Prum's work is helpful in its characterization of the forces of sexual selection and natural selection, in this chapter I argue that it fails as a theory of art. In particular, his account fails to exclude the possibility that these behaviors are best understood as language-like, which provides support for the account I have presented here. Although sexual selection may lead to traits that we judge to be beautiful, this is not sufficient to say that they are art. At the same time, this does not rule out the possibility that some forms of bodily adornment should be considered art—but this cannot be merely in virtue of their perceived beauty.

Prum's account of co-evolution helps us illuminate further some questions about the distinction between natural and non-natural meaning that may be lingering from my initial characterization of Grice. I close the book by considering what Prum's notion of co-evolution means for those who see evolution as implying that the natural world is without a designer.

A Personal Theory

Before I turn to consider previous attempts to treat bodily adornment as a language in the next chapter let me first make some further remarks about the subject of this book. In her 2016 book *The Minority Body: A Theory of Disability* philosopher Elizabeth Barnes writes that for her the subject matter is personal, given that she has a condition that is classified as a disability (Barnes 2016). She also notes that in a way it is personal for those who are not disabled as well, because being classified as "not disabled" is something that makes up their identity (Barnes 2016).

In the same way, this book is personal to us all. It is personal to me because I have a body with which I move about the world, and the way my life has gone has been shaped in certain ways by how others have perceived my body. I have been seen as a threat or not a threat. I have been let into a nightclub or turned away. The same is true for you as well. Whenever I discuss the material in this book with others they always see it through the lens of their body, their experience, what they wear, how others respond to them. And of course, anyone would.

We all have unique bodies and unique experiences of moving through the world in them. We also can consider what it would be like to have another body, to wear other types of bodily adornment—or perhaps more importantly to be the sort of person who wears other types of adornments, is perceived differently. My perspective on the topic of bodies and bodily adornment can't help but be colored by my experience in my body. I hear about, read about, and ask questions about the experience of having another type of body but never can know what that experience is actually like. I will give voice to some philosophers and authors who describe their experience in a range of different types of bodies in later chapters, but recognize that I am limited in my ability to fully understand their experience—or your unique experience moving through the world in your body.

We do not view all bodies the same way. We do not view adornment the same in all contexts either. As I will discuss, adornment does not mean things in a vacuum—it means something on a *particular* body.

It can be uncomfortable to think about bodies, to acknowledge that we have bodies, and to consider the ways that our bodies are perceived by others. Perhaps it is easier to think of ourselves like Descartes did—in terms of minds, souls, or brains in a vat, the "ghost in the machine." For many people, thinking about their bodies causes them to feel shame. This is part of what our bodies have come to mean to us, through some lifelong process that none of us share exactly in common (Taylor 2018; Johnson Forthcoming).

A Word on Fashion

I should also note that the issues I discuss here are certainly not limited to those concerns of the subset of the population who consider themselves to be interested in fashion. In a way what I will discuss is much more mundane, more everyday. It applies as much to a model at a Chanel show as it does to a philosopher who "throws something on" to attend a lecture. Just as philosophy can be subject to elitist gatekeeping, so can the subject of fashion.

Some people might feel that because they dress simply that meaning in adornment does not apply to them. In the afterword to his 2020 book *Adornment* philosopher Stephen Davies writes,

> As a male of the baby-boomer generation, I'm a poor advertisement for this book. I don't wear jewelry or makeup. I'm bereft of tattoos and piercings. I've been accused of dressing like a homeless person. I never wear suits, and if I did, I wouldn't put a flower in the lapel. So, why did I choose to write on adornment? (Davies 2020: 207)

Davies goes on to explain that he became interested in the subject of adornment through his other work on beauty. But Davies is not outside the relevant realm of bodily adornment. A "male of the baby-boomer generation" has their own relationship to their body and bodily adornment, just as much as anyone else.[1]

We know that what we wear is interpreted, and sometimes are given feedback about how. We choose to get dressed the next time well aware of the feedback we have been given in the past. Whether or not we choose to change what we wear at different points in our lives reflects a number of things, including, perhaps what we are able to get away with, and our place in the social structures around us.

What Meanings Do

As a starting point to consider meaning in bodies and bodily adornment I will present a number of cases in the next chapter. We have already seen the house finch and in the later chapters we will consider a queen's crown, an ironic sweater, a black armband worn in protest, and numerous other cases. There are different mechanisms behind these types of meanings I will present.

Let me end this chapter with a general comment about what meanings do. Some meanings co-evolved with their interpreters because they serve a purpose—sometimes we believe things because we recognize someone else intends us to believe something. Sometimes we believe things because we are presented with a state of the world that provides direct evidence. These two ways of meaning will be broken down and explained in terms of H. P. Grice's distinction between natural and non-natural meaning in the following chapter. I will argue that the differences between these two reasons to believe play a crucial explanatory role in understanding the two basic ways we communicate with bodies and bodily adornment. In the final chapter I will connect this discussion with Prum's account of the co-evolution of beauty. As we will see, there are important connections to be drawn out.

Ultimately what interpreting meaning does is change our beliefs about the world. When someone means something they provide reasons to continue to hold our beliefs or reasons to change our beliefs. A female house finch viewing a dull red male or a bright-red male has a resulting mental state that affects her behavior. The red feathers have different meanings that cause her to act in different ways. Meanings are a part of inferences we make about how to act in the world. This is why we *care* what things mean; interpreting meanings leads to action.

2

Taking Adornment Seriously: Structuralism and Meaning

Introduction

We adorn the body for a number of reasons. There are, of course, the practical reasons. We put on a warm winter coat because it protects us from the cold. We put on a hat because it blocks out the sun. Bodily adornment also has a certain phenomenology for the wearer. It is difficult to imagine having a perfect day in a too-tight polyester shirt. We might feel ourselves relax when we come home and change out of work clothes, taking off our blazer and putting on a comfortable sweater.

In addition to these practical and phenomenological reasons for wearing what we do, we use our clothing to communicate. In this chapter I will begin by clarifying the metaphysical boundaries of my object of inquiry: adornment. I will next introduce some specific cases of meaning in bodily adornment. I will then present and consider previous attempts by philosophers to take meaning in bodily adornment seriously. Finally, I will introduce the Gricean account of meaning before going on to provide details in the following chapter.

Adornment: The Object of Inquiry

I would first like to clarify what I take to be the object of inquiry. I have stated that this project is one of considering the communicative power of bodies and bodily adornment. By using the term "bodily adornment," I mean to include any additions to the natural form, permanent or non-permanent, from toupees to hair plugs, mascara to eyelash extensions, school uniforms to tattoos. I shall often use the terms "dress" or "clothing" when I happen to be discussing that particular form of adornment but this should usually be taken as shorthand for the broader category of bodily adornment.

The Functions of Adornment

My focus on the communicative or intentional potential of dress in this book does not entail that dress does not have other purposes or that articles of adornment may only be a means of communication. Adornment has both higher and lower aspirations than those I focus on here. It certainly can have a functional purpose, such as keeping off rain or keeping one warm—and sometimes there is tension between the functional and beautifying aspects of some piece, as I will address later on in my discussion of Darwin and the peacock's tail.

Adornment can have aesthetic aims and sometimes these elevate fashion to an art (Baxter-Wright 2012; Svendsen 2006: 90–110). Many fashion designers consider themselves to be artists. In 1913 Parisian fashion designer Paul Poiret, who was instrumental in bringing about the end of the corset, said, "I am an artist, not a dressmaker" (Svendsen 2006: 91; Koda and Bolton 2007: 150). Similarly, some of fashion designer Elsa Schiaparelli's fanciful designs were created in collaborations with artists including Salvador Dalí and Meret Oppenheim, who worked in her shop (Baxter-Wright 2012: 74; Wood 2007). These items of adornment "were probably just as advanced as most Surrealist pictorial art produced in the 1920s and 1930s" (Svendsen 2006: 107; Figure 1). I will discuss the links between adornment and art in detail in the final chapter.

One might insist that these artistic qualities of dress mean it can never be treated by theories from philosophy of language. However, poetry and much prose also have similar artistic qualities and they are treated as a part of philosophy of language. The fact that dress may have additional practical applications or artistic qualities does not preclude it from being treated within philosophy of language.

Attention to the higher and lower aspirations of dress is compatible with the position I advocate, which allows for the other roles dress plays besides those communicative roles that are my primary focus here. I am not making the claim that communication is the *only* thing we do with clothing, but one of the important things, and one that has been neglected.

Figure 1 Women's hat designed by Elsa Schiaparelli, 1949. This red satin Schiaparelli hat with peephole and diamond clip from Van Cleef & Arpels bears the hallmarks of surrealist art. *Source*: Getty Images.

Adornment and Fashion

Such a starting point raises questions about the relation of this project to fashion. Colloquially the term "fashion" is used in a number of ways. One who says she "works in fashion" means she works designing or marketing clothing. A "fashion magazine" has a focus on new high-end clothing items and all the trappings of what are deemed to be a glamorous lifestyle, from where one ought to vacation to what one ought to eat for lunch.

Moving past these ordinary language uses, philosophers have defined "fashion" in a number of different ways (Hollander 1993; Hollander 1994; Svendsen 2006; Pappas 2008; Farennikova and Prinz 2011; Pappas 2015, 2017). Some define it quite broadly. Lars Svendsen writes that "Fashion is not just a matter of clothes but can just as well be considered as a mechanism or an ideology that applies to almost every conceivable area of the modern world" (Svendsen 2006: 11). Others, such as Gilles Lipovetsky, characterize fashion in terms of a process of change and novelty—a product of advancements in production of our modern era (Pappas 2008: 5; Svendsen 2006: 21–35). Fashion has also been understood as a project of mass imitation (Pappas 2008: 5). Others still limit fashion to a subset of clothing that meets some aesthetic standard. Anne Hollander writes that fashion is comprised of "the entire spectrum of attractive clothes styles at any given time" (Svendsen 2006: 11). Understood in this sense, calling something "fashion" would be a judgment of its success, akin to deeming a sculpture to be "art."

It is because of the compelling arguments on all sides of the philosophical debate on this topic and the complexities in our ordinary use of the word "fashion" that I will henceforth do my best to steer clear of the term in this book. My object of inquiry can be seen in some ways as narrower than some conceptions of "fashion" and in other ways broader. It could be considered narrower because I will not consider cases of "fashion" that are not bodily adornment; it could be considered broader because neither aesthetic success nor newness is a requirement of the class of cases I will consider. Although I will draw on the work of a number of theorists who see themselves as operating in the realm of fashion, I will use their arguments toward what I see as the more agnostic category of bodily adornment that I discuss here.

"Adornment"

"Adornment" has, of course, at its core the word "adorn"; it picks out an action—drawing on a verb rather than a noun. Something becomes adornment because it plays a certain role in our lives. In his 2020 book *Adornment: What Self-Decoration Tell Us about Who We Are*, philosopher Stephen Davies also uses the term "adornment" (Davies 2020). Davies has long worked at the intersection of philosophy, anthropology, and archaeology, where the use of the term "adornment" is frequent. "Adornment" does not carry the cultural-specific baggage of a word like "fashion" and is broader than a term like "clothing" which might not have the range to cover cases spanning the globe and at different points in human history.

In his book Davies presents the thesis that adornment, as he understands it, results from a defining behavior of our species. As he puts it, "adornment is not merely apparent in every culture, but that it is nearer to being universal—something every person does—than any other human behavior" (Davies 2020: 3). He supports this thesis with a number of concrete examples given throughout the text.

As was the case with the term "fashion," as used by a number of theorists, Davies at times excludes things I classify as adornment and at other times includes those I wish to exclude. Explaining this is a helpful way to set the stage by clarifying my own view on adornment—which at times is in contrast to Davies's.

Davies begins by defining the term "adornment"; he includes not just the way we decorate ourselves but also the way individuals decorate their "possessions, their living quarters, and their wider environment" (2020: 2). In this way Davies uses the term broadly to mean what I would describe as "decoration," a term he also uses sometimes in the text. There is a history of the term "decoration" in the anthropology literature too (e.g., Strathern and Strathern 1971). But it is not clear what Davies gains by understanding "adornment" so broadly, to include possessions and homes. Indeed, after presenting it in this broad way in the first couple of chapters, he then goes on to focus on adornment of the self in particular in the majority of the book.

On my understanding adornment must be adornment of the *body*. It is a metaphysical category of objects that are individuated by their role in bodily behavior. I agree that there may be parallels in certain ways with decorating the home but these are importantly distinct categories in my view.[1] Bodies have features that lead to us being categorized into genders and races that are not present in other things we decorate. Bodies also have phenomenology. Because of the important unique traits of *bodily* adornment I carve a metaphysical space between bodily adornment or "self-decoration" and decoration of other things like our homes.

Davies also includes a limiting condition on his understanding of the term "adornment" that excludes certain forms of *bodily* adornment. Adornment, for Davies, requires "beautifying or other aesthetic intentions" (Davies 2020: 13). This condition serves to narrow adornment beyond what we wear. Certainly some adornment—as I understand and will use the term—does have this intention, but not all. On the contrary, sometimes with adornment we wish to blend in, to be weird, to elude a sexual gaze, or to shock, anger, or even repulse (Johnson 2021). As I understand "adornment" it encompasses all the possible intentions we have when we adorn; there is no requirement that this be to beautify or aesthetic in nature (although depending on how broadly we understand aesthetic maybe this covers a great deal). Our aims with adorning are complex (Johnson 2021). Throughout the book I will discuss a number of intentions one has with dressing, including very specific communicative intentions, as well as broader non-aesthetic intentions such as to convey power and subjugate.

The Explananda

Let me now present some of the complex language-like ways we communicate with bodies and bodily adornment. As French philosopher Roland Barthes wrote in an early preface to his 1967 book *The Fashion System*, "Clothing is undeniably meaningful: humans communicate via clothes, tell each other they are getting married, being buried, going hunting or to the beach, if they are department store staff or intellectuals, if they are doing their military service or painting" (Barthes 2005: 77). I see these types of examples as the explananda that a theory of communication by bodily adornment ought to account for. These cases demonstrate that communication with bodies and bodily adornment is far from simple.

First, statements of dress can pertain to the present time, the past, or the future, as in (1–3).

1) This is a special occasion (present).
2) I won a Purple Heart in the military (past).
3) I'm going to prom later (future).

Second, clothing statements seem to be primarily about oneself, as in (4), but can also be directed at others as in (5). They can signal social group membership as in (6):

4) I am a police officer (self-directed).
5) I'm trying to impress you (other-directed).
6) I am a mod (group-directed).

Third, statements of dress can be about "here," as well as places one has been, or is going (7).

7) I just came back from the Bahamas (distal location).

Fourth, dress can even be a part of complex ways of referring such as by what is called "implicit reference" (Neale 2007). Some nouns such as "mayor" must be tied to some place or other; no one can hold the title of "The Mayor"—they must be "The Mayor" of somewhere, such as "The Mayor of New York City" and in virtue of this feature a speaker uttering such a noun is said to implicitly refer to some location (Neale 2007). Similarly, a woman wearing a crown is understood to be a queen. As with "The Mayor," "The Queen" is the sort of thing one can be only with respect to some place or other. This means that with this item of dress—the

Figure 2 Lorena, Paul, and Mary Beth Tinker. Des Moines students Paul and Mary Beth Tinker wearing the black armbands that led to the *Tinker v. Des Moines* 1969 Supreme Court case. Their mother Lorena is pictured on the left. *Source*: Getty Images.

crown—the wearer implicitly refers to some place or other that she is the Queen of. The wearing of a crown (8–10) can convey the following:

8) I am the Queen [of England] (implicit reference).
9) I am the Queen [of Sweden] (implicit reference).
10) I am the Queen [of the Mahtomedi High School Prom] (implicit reference).

Adornment as Protected "Speech"

Adornment is protected speech according to the U.S. Supreme Court. An act of meaning in the form of a black armband worn in protest of the Vietnam War (11; Figure 2) was memorialized in a famous Supreme Court Case, *Tinker v. Des Moines School District*, 1969. In this case the court ruled that three students who had chosen to wear black armbands to protest the Vietnam War were protected by the Free Speech Clause of the First Amendment (*Tinker v. Des Moines School District*; for other important rulings on adornment, see also *Pugsley v. Sellmeyer* 1923; *Ferrell v. Dallas School District*, 1968). In this case it conveyed:

11) I object to the war in Vietnam (attitude toward event in distal location).

The convention that remains today for one to wear a black armband as a signal of mourning originated with the black clothes widows would wear in Victorian times (Hollander 1993). As I will discuss in detail in the next chapter, this item of bodily adornment has a decades-long history that we can track over time. A black armband—the same garment—can be worn to express a different meaning in different contexts.

Communications made through dress are not restricted to being simple statements about the wearer. Examples 1–11 show that dress can be used to communicate a very wide range of complex propositions. But how?

Dress and the Code Model

There have been previous attempts made to explain clothing as something akin to language, some more successful than others. In *The Language of Clothes* Alison Lurie (1981) uses the starting point of clothing as language as a metaphor to lead into her insightful social history of fashion. This work provides a rich discussion of social themes in fashion history and I will draw on Lurie's keen observations at times throughout the book but her treatment of clothing as language is rather thin. I agree with philosopher Lars Svendsen's evaluation that her "analogies are not particularly convincing, and the highly direct manner in which Lurie interprets everything often comes close to unintentional parody, as when she claims that a tie in bright colors expresses virility, or that a clergyman without a tie has been 'symbolically castrated'" (Svendsen 2006: 65). To make progress on the issues I have laid out here we require a more robust treatment of clothing as a language that goes beyond metaphor and analogy.

The most serious attempt to systematically study meaning in adornment was undertaken by the French philosopher Roland Barthes in the 1950s and 1960s. Barthes's project failed in ways he recognized, but considering his methodology and *why* it failed proves instructive. It also sheds light on broader problems with structuralism, the framework Barthes worked within. These issues for structuralism will again become apparent in my seventh chapter, on prehistoric bodily adornment.

Barthes's ambitious project aimed at identifying "the meaning of clothing" as he understood it. His project is exhaustive and is considered to be "the most theoretically ambitious attempt to consider clothes as a kind of language"

(Svendsen 2006: 66). It is for this reason that I give Barthes much of my attention in this chapter. Identifying the problems with Barthes's attempt helps flag which changes in approach would be superior and can serve to point us in the right direction.

Barthes's objective in *The Fashion System* (1983), first published in French as *Système de la mode* in 1967, and articles later collected and published in English as *The Language of Fashion* (2005) was to identify the fixed meanings of clothing that hold across different occasions. Toward this end, Barthes engaged in a sort of field research survey of fashion magazines, *Elle* and *Jardin des Modes* from June 1958 to June 1959. At the time these magazines included full-page themed editorial photo shoots accompanied by a brief description of the garments and what was conveyed by them. Barthes used this juxtaposition of image and text description to identify correlations between aspects of clothing and words in the corresponding description of the ensemble.

Barthes saw this as his most sustained exercise of applied semiology, the "science of signs" which was first developed by Swiss linguist Ferdinand de Saussure in the early twentieth century (Barthes 2005: 70–71). Barthes saw in the way clothes are described in fashion magazines with wording such as "the accessory makes springtime" and "this women's suit has a young and slink look" a link between object, "a signifier," and its meanings, "a signified" (Barthes 2005: 41). He writes,

> It is as if I was being offered, simultaneously, a text and its glossary of words; all I will have to do (in theory) is start from the signs in order to define straightaway the signifiers: defining them is basically to isolate them. . . . Then, all I need to do to understand the all-signifying structure of the clothes is look within each unit for the aspects whose opposition helps create meaning (blue/red? blue/white? brooch/flower? camellia/rose?). (Barthes 2005: 43)

Barthes's research project was one of identifying these "links" between signifier ("a woman's suit, a pleat, a clip brooch, gilt buttons, etc.") and signified ("romantic, nonchalant, cocktail party, countryside, skiing, feminine youth, etc.") (Barthes 2005: 42).

It is, of course, no small task to isolate which particular aspects of a piece are responsible for some specific attributed property. And even if it were possible in one case to, say, attribute the elegance of some particular dress to its striking slate blue color, this same blue color when on a lamé hot pant will not endow the hot pants with the attribute of elegance. Any aspect that has some property in one context can fail to have it in another context. Surely in all of these cases the success of the object in question cannot be straightforwardly attributed to any single part, but to their effects in unison.

Barthes himself acknowledged a number of the problems his project faced, writing, "Unfortunately, the fashion magazine very often gives me links where the signifier is purely graphic (*this* nonchalant ladies' suit, *this* elegant dress, *the* casual two-piece); I then do not have any way—unless intuitively—to decide just what in *this* suit, in *this* dress or in *the* two-piece signifies nonchalance, elegance or casualness" (Barthes 2005: 44). He adds, "If I read that a *square-necked, white silk sweater is very smart,* it is impossible for me to say . . . which of these four features (sweater, silk, white, square neck) act as signifiers for the concept *smart*" (Barthes 2005: 70). Despite these obstacles, Barthes pressed on with the project, although he ultimately described it as "precarious," and concluded that "we have nothing but doubts (and often unpleasant certainties) about how well it has been fulfilled here" (Barthes 2005: 70, 81). Lars Svendsen writes that "even at the time of publication Barthes considered it in some ways a failure" (Svendsen 2006: 66) and suggests that "it is no coincidence that it was his work on *The Fashion System* book that led Roland Barthes to abandon classic structuralism" (Svendsen 2006: 71). It is after this failed project that Barthes wrote his famous piece "The Death of the Author," advocating for an entirely different theory of meaning (Barthes 1978).

As he recognized, Barthes's project was not successful in meeting its stated aim to identify a complete "Fashion System" in the semiotic tradition. This result is exactly what we would expect if dress is a complex linguistic device with many of the nuances of language. Barthes's project treated dress as though it can be fully explained by a system of fixed codes—correlations between a signifier and a signified.

This sort of model is very limited. Anthropologist Dan Sperber and linguist Deirdre Wilson, in their 1986 book *Relevance: Communication and Cognition*, conclude that if we treat verbal utterances as codes we run into the same problems (Sperber and Wilson 1986). In such attempts "all that can be reasonably maintained is that the use of a code plays some role in this particular communication process, without perhaps fully explaining it" (Sperber and Wilson 1986: 27). An adequate explanation of meaning in dress must go beyond the code model to a speaker-based conception of meaning in dress. In other words, we can take the result of Barthes's project as evidence that in clothing, as in speech, "a coding-decoding process is subservient to a Gricean inferential process" (Sperber and Wilson 1986: 27). One of the main problems with a structuralist view is that language does not operate like a code, with inputs and outputs.

In the first chapter of their 1986 book *Relevance*, Sperber and Wilson motivate a Gricean-type view by contrasting it with a structuralist sort of code model. In a code model communication is understood as a matter of one agent producing an output, which is interpreted by another agent. A sender sends a signal to a receiver. These theories are based in information theory—a tradition that began with radio communication in work by Claude Shannon and continues today through the work of people like Brian Skyrms (Shannon 1971; Skyrms 2010; Godfrey-Smith 2012).

Treating dress not as a sort of code or signal but as the sort of thing that has meaning *in virtue of being a means by which people mean things*, we will have more success. The intention-based model of meaning in clothing called for is broadly Gricean—that is, it follows the basic processes outlined by philosopher Paul Grice in his theory of meaning (Grice 1957, 1989).[2]

A More Nuanced Account of Meaning

The structuralist account of meaning presented by Roland Barthes fails as a theory of how we communicate through bodily adornment. Barthes's attempt fails for the same reason many theories of linguistic meaning failed: they treat meaning as the result of a system of codes—an assumption that leads to theories that can never fully explain communication (Sperber and Wilson 1986). Consideration of the reasons Barthes failed provides support for a more nuanced account of meaning.

Although I disagreed with his semiotic approach, let us assume that we agree with the assertion in the Barthes quote that started the chapter: "Clothing is undeniably meaningful." What does this claim amount to? Where does the meaning come from? Is clothing meaningful because we decided it is meaningful or does it have meaning on its own? In presenting an account of meaning in bodies, a distinction must be made between two different types of meaning: natural meaning and non-natural meaning. As I will explain, the distinction between natural meaning and non-natural meaning is a metaphysical one, but it has epistemic implications.

In his canonical 1957 paper "Meaning" Grice distinguishes between natural meaning and non-natural meaning. Natural meaning is, roughly, meaning things in the world have, such as in "Those clouds mean rain" or "Those spots mean measles." This is contrasted with what he calls non-natural meaning, meaning that we give and interpret on the basis of communicative intentions.

Natural meaning, or meaning$_N$, applies to bodies in the sense that we take certain bodies to convey information about the person who has this body. We believe taller people are better leaders. We categorize people into gender and racial categories. In this sense we take bodies themselves to have meaning—Gricean natural meaning. Natural meaning is also found in the animal world. I will present a number of illustrative examples in later chapters. If we change the ways our bodies look—which we do with adornment—we can thus change the associations others have with our bodies.

Natural meaning is the sort captured by saying that "spots mean measles" or "the budget means we shall have a hard year" (Grice 1957). In such cases "x means p" can be restated "the fact that x means p" without a change in meaning. Non-natural meaning, or meaning$_{NN}$, in contrast, is the sort of meaning captured by saying "those three rings of the bell mean that the bus is full" (Grice 1957). In such cases "x means p" can be restated "by x so-and-so means 'p'" without a change in meaning. The important point of demarcation between natural and non-natural meaning for Grice is that non-natural meaning is *intentional* and is the sort of thing that *somebody* means by some act. It will not always be obvious if something is natural or non-natural meaning, but Grice says in most cases we will lean one way or the other (Grice 1989: 215).

Grice explains in great detail what makes it the case that some instance of non-natural meaning has the meaning it has: some utterance, x, means some content, r, because some utterer, A, uttered x with certain meaning-intentions with respect to the utterance. The basis of non-natural meaning on the one hand is not causal; it is intentional. We can say, "A non-naturally meant something by x" is roughly equivalent to "A uttered x with the intention of inducing a belief by means of the recognition of this intention" (Grice 1957). Natural meaning, on the other hand, is not intentional. With natural meaning it is not some A that means something by x, but simply some x that means something.

Natural Meaning and Imitation of Natural Meaning

Let me now give another example of natural meaning to help further illustrate the difference between this and non-natural meaning.[3] In graduate school at the City University of New York I used to teach a winter-term course at City College. I would take the subway north to get to campus, walk up a steep hill, and teach in an old classroom with chalk boards—my first and last time teaching with chalk rather than on a whiteboard. It was January in New York and like most of my fellow train riders I was usually wearing black. My first few times riding home from City College I would sort of take a breath and come down from my thoughts of teaching to take in my surroundings. Looking down, I saw my clothes were no longer black. It turned out I was absolutely covered in chalk. In fact I had a line around my hips where they hit the ledge on the bottom of the board. I tried to rub it off, but would only rub it in, adding whatever chalk had been on my hands. It was futile. It was clear I had just been teaching. The evidence was covering me.

My formerly black clothes now covered in chalk had meaning—natural meaning. The state of my clothes meant I had been teaching. It's true that not everyone would be able to interpret this case of natural meaning, but surely my fellow professors—who have experienced the same thing—would. The type of meaning seen in my chalk-covered self parallels Grice's example of clouds meaning rain. Our clothes and other forms of adornment or visible bodily changes can provide direct evidence of where we have been or where we are going.

We are so good at making these sorts of inferences that we might not realize we are making them. If a friend is dripping in sweat after the gym we will assume they had a hard workout. If a job candidate's shirt is covered in sweat during an interview we will assume they are nervous. In these inferences the surrounding facts are relevant and used to assess the situation. This may or may not be conscious.

What is important with these types of meaning is that the intention is *not relevant*. That is, we come to take the body or bodily adornment to have natural meaning not because the other person wants us to but because we simply take some state of the person (e.g., chalk on clothes, sweaty) within the conditions of the state of the world (e.g., on the subway, after a workout, during an interview) to *mean* something. Here, the thing itself has meaning. In interpreting these things we make no attributions about what somebody meant, and do not need to consider intentions.

In this way, with natural meaning, our more base, animalistic selves as bodies—as organisms that are themselves indicators of things to those around us—take center stage. This is no coincidence, and Darwin himself was fascinated by the parallels between bodily "expression" in man and animals. Animals assess each other using various indicators, some that we share and some that are unique to each species. I will come back to this point in later chapters but let me just make a brief comment about it now.

Some animal markings can be a direct result of the ability to navigate their environment. As mentioned in Chapter 1, the red color of male house finches is believed to be a result of their diet, and those who have an insufficient diet are yellow instead of red (Tekiela 2000). Thus, the degree to which a male house finch has red coloring can be taken as a direct indicator of his ability to find proper sources of food, a skill of paramount importance to any potential mate. This is just one of the many examples of natural meaning in the animal world.

Humans have, of course, evolved a number of similar signaling mechanisms by the same processes. Our unconscious facial movements, for example, are a valuable transmitter of information to others. The ability to express submission to a threat through automatic facial movements can be of great benefit because it may cause the aggressor to relent (Choi et al. 2004: 314). I will say more about the expression of emotions in Chapter 8. The strong contrast between the whites of our eyes and our irises is also a great adaptive advantage because it provides the ability for others to track our gaze and see what we are attending to (Tomasello 2010). This helps us to think about the minds of others.

Structuralism versus Intentionalism

Grice's theory is grounded in this capacity to think about the minds of others. His account of non-natural meaning is importantly different from the sort of explanation found in the structuralist literature, although both have similar explananda. Structuralism requires establishing regularities between a signifier and a signified. For Grice it is not this regularity that is important but a speaker's intention when producing some utterance. With Gricean non-natural meaning we do not explain meaning in terms of a mere tendency for some act or utterance to be correlated with some meaning. As competent speakers of languages we recognize that what somebody means by an utterance often goes far beyond the "coded" meanings of the words themselves.

The crucial point about this discussion is that with the Gricean position the bearer of meaning moves beyond the words themselves to consider the speaker or utterer of those words. When we interpret actions with non-natural meaning we are not decoding signifiers, we are looking to identify what somebody else is aiming to get across. This is a significant shift from structuralism in the understanding of what the focus of interpretation is, and in what our ultimate objective in communicating is.

Black Armbands of Mourning

Upon reflection our capacity to recognize the intended meanings of adornment is remarkable. As presented earlier an item of dress can refer to and express a negative opinion of an event in another country. This was seen with,

11)   I object to the war in Vietnam (genuine mental attitude toward event in distal location).

Here adornment was seen as protected speech in the 1969 Supreme Court case *Tinker vs. Des Moines School District*.

The television show *Fawlty Towers* provides us with another, highly complex instance of meaning with a black armband. In a memorable scenario, Manuel, an incompetent employee, has a pet rat that he believes to be a rare Siberian hamster. The equally incompetent hotel owner, Basil Fawlty, has banned the rat, which is a violation of health code, from the hotel. Manuel wants to keep the rat and protests, but Basil insists he get rid of it. Manuel pretends to reluctantly obey Basil but in fact hides the rat in the shed. In the next scene Manuel walks into the kitchen wearing a black armband. As soon as we see Manuel we know what he means by wearing this black armband. The black armband means "I am mourning the loss of my hamster" (27).

12)   I am mourning the loss of my hamster (feigned mental attitude about action taken by person present).

Although a viewer can instantly interpret the black armband, this is an incredibly complex act of meaning.

This case surely is a novel instance of a black armband being worn to mean "I am mourning the loss of my hamster." However, with non-natural meaning in bodily adornment we are capable of making and interpreting novel meanings. We also know that it would not mean "I am mourning the loss of my rat" because Manuel is the one who is wearing the armband and he does not believe that the creature is a rat. Further, we are watching a fiction and can recognize that although we, the viewer, know that Manuel is using this armband deceptively and is communicating something false (he is not mourning the loss of the rat because it is hidden in the shed) we at the same time recognize that Basil is not aware of this and that he might think that Manuel is really in mourning. All of these levels of interpretation and recognition happen in us very quickly as we watch the scenes play out.

The convention that one wears an armband as an act of protest is quite common, and would be recognized by most viewers of the item. However, even with this recognition, the viewer must know something of the wearer's intentions to work out what it is that is being protested; this is not something that can be worked out from knowing the general convention about armbands. It certainly is not a part of the armband itself that *it* means "I am mourning the loss of my hamster." Indeed, Manuel's use of an armband to communicate this is likely a novel use. In this way, an armband is analogous to some words such as "he" that have meaning that are partly a function of conventions—that one be speaking of a male person—and partly a function of the speaker using this word on a certain occasion with a certain intention. To explain why on one occasion some item of bodily adornment has a certain meaning we need to look beyond the signal itself and appeal to facts about the person who wore the item. Here it makes the most sense to not speak of the *armband* meaning something, but that *Manuel* means something by wearing it on that occasion, with that intention. The conventions do play a role here but they are not enough to individuate what is being protested in this case. I will say more about how we should understand the role of conventions in the following chapter.

The Bulldogs

With adornment, as with linguistic utterances, there is also the potential for misunderstanding. Of course, wearers do not always gauge their interpreters correctly, and may assume a shared background that is not there. Being misconstrued and taken to be literal when one is being ironic is one thing, but sometimes the consequences of being misinterpreted can be more dire.

A deadly misunderstanding was seen in a real-world case detailed in the *New York Times* in 2013 (Wollan 2013). As the article describes, beginning in the 1980s, a small college team in Southern California experienced a tenfold spike in sales of red jerseys and merchandise bearing the insignia of its team's bulldogs mascot. The reason was that a local gang also named "the bulldogs" had adopted the colors and jersey as a part of their ensemble to signify gang membership (Wollan 2013). In 2006 the bulldogs gang was responsible for 70 percent of the city's shootings (Wollan 2013). The team mascot—a macho-looking bulldog with a studded collar—was well suited to capture the ethos of both the football team and the gang.

However, in such an environment, there surely are stark differences bween the consequences of communicating that one is a fan of the local sports team and communicating that one is a member of the most violent gang in the city. Unfortunately, because both contents were conveyed using the same item of dress there was potential for the wearer's intentions to be misconstrued. In 2011, confusion caused by the shared mascot came to a head when Stephen Maciel, a father of four, who, according to the police, was not affiliated with any gang, was shot and killed in the parking lot of a liquor store while wearing a bulldogs shirt (Wollan 2013).

Communications made through dress are not restricted to being simple statements about the wearer. These examples show that dress can be used to communicate a wide range of complex contents. Speaker meaning is the actualization of one of some signal's potential meanings—what *somebody* means by some word or item of adornment on some occasion. It is because of the complexities brought to light by cases like the armband, and the bulldogs jersey that it becomes even more clear that a structuralist account of meaning can never be fully explanatory. We need the nuance of the Gricean account.

The Requirements for Natural Meaning

This doesn't mean, however, that the distinction between natural and non-natural meaning is clear, obvious, or neat. Grice does not spend much time detailing what it is that makes it the case that instances of natural meaning have the meaning they do—rather, he uses a somewhat broad notion of natural meaning to clarify his notion of non-natural meaning against.

The form of the human figure naturally means certain things about the person—things about, say, their age, level of health, and how good of a mate they would be. These features affect how the individual is perceived by others. For instance, psychologists have demonstrated that taller people are considered to be more persuasive, more attractive as mates, and more likely to emerge as leaders (Judge and Cable 2004). Height is also positively related to income. One study found that "an individual who is 72 in. tall would be predicted to earn almost $166,000 more across a 30-year career than an individual who is 65 in. tall" (Judge and Cable 2004: 437). The taller presidential candidate has won the U.S. election two-thirds of the time (Maclean 2019).

Facial features, too, are thought to be indicators of personality. In a 2005 study of faces Leslie Zebrowitz and Joann Montepare found that "babyfacedness"—

which was indicated by features such as "a round face, large eyes, small nose, high forehead, and small chin"—was correlated with judgments of incompetence (Zebrowitz and Montepare 2005). They explained the findings in the following way:

> According to the ecological theory of social perception, our ability to detect the attributes of age, health, identity, and emotion has evolutionary and social value. Thus, we have a strong, built-in, predisposition to respond to facial qualities that reveal these characteristics. . . . In this case, our impressions of babies (submissive, naïve, and weak) are extended to babyfaced adults who are consequently perceived as less competent than their more mature-faced peers. (Zebrowitz and Montepare 2005: 1565)

As with all such studies, correlation can be demonstrated but causation is very difficult to establish. For instance, it has been demonstrated that attractive people are happier (Diener 1995). But is this because being happy makes them attractive? Does being attractive make them happy? Is it a bit of both? Further complicating the matter, Ed Diener and colleagues found that happy people do more to enhance their beauty (Diener et al. 1995). This, of course, does not rule out the possibility that the causation goes in other directions as well.

It is not just the outside perception of others that is at play here either. Happy people perceive themselves to be more attractive (Diener et al. 1995: 128). This means that perceived attractiveness could also enhance self-esteem (Diener et al. 1995: 128), and then this could in turn increase the ability to achieve goals that would lead to more happiness (Judge et al. 2005).

Natural Meaning and Truth

Let me include here some comments on natural meaning and factivity. You might here be thinking of the fact that not all tall people are good leaders and not all babyfaced people are incompetent. For Grice natural meaning is factive. That is, with natural meaning, something cannot mean$_N$ p if p is in fact not the case. Those clouds cannot mean rain if it will, in fact, not rain. This means that natural meaning is quite different from what we understand to be evidence that something is the case. Evidence need not be factive; natural meaning needs to be.

Natural meaning's value as an epistemic concept is limited, but Grice did not set out to develop an epistemic theory. His aim was to develop an account of *meaning* and it is right of him to have said that those clouds cannot mean rain if it does not in fact rain. The clouds might seem to mean or indicate or suggest or increase the probability that it will rain, but none of these are *meaning*, and it is meaning that Grice wanted to get at. To me, Grice's account of the metaphysics of meaning is quite intuitive, but not all agree. Philosopher Arda Denkel, in a book entitled *The Natural Background of Meaning*, writes the following:

> Pace Grice, it seems perfectly acceptable to say "These spots mean measles, but actually she has not got the measles", or "Though the present budget means that we shall have a hard year, in all probability we shan't have". That these spots mean measles does not entail that the bearer of them has measles, and that this budget means a hard year ahead does not entail that the year will actually be hard. It seems to be easily conceivable that by injecting a chemical extract we could cause spots on one's skin, indiscernible in quality from those of actual measles. Such spots would deceive a physician. It also appears possible (and in many countries highly probable) that halfway through the year the budget will be abandoned, and another with less stringent measures adopted. (Denkel 1998: 93–94)

Denkel in this quoted passage says something that seems to clearly be false. How could it be the case that "These spots mean measles, but actually she has not got the measles"? How could it be true that "Though the present budget means that we shall have a hard year, in all probability we shan't have"? As I read them these both consist of a baffling contradiction.

We could certainly say, "These types of spots typically mean measles, but actually she has not got the measles," but that *these spots*: that is, these very spots before us, these *particular spots*, mean measles is something different. I can only understand Denkel to have had in mind that these *types of spots typically* mean measles—to be using the term "means" in a misguided epistemic sense.

There are constraints on what something can mean. For natural meaning this is limited to factivity; for non-natural meaning this is limited to what the speaker intended. Meaning for Grice is a metaphysical concept, *not* an epistemic one. Now, we as beings who go about the world encountering people and things are not able to divine metaphysical truths. We live in the world of the epistemic. It might be the case that those clouds mean rain, but, of course, we can never know this with 100 percent certainty. To treat Grice's account of meaning$_N$ as an epistemic theory and then fault it for not being a very good one would be to misinterpret his aims.

Natural Meaning and Bodies

Recognition of the limits of our capacity to interpret natural meaning raises a number of questions. When we consider the natural meaning that bodies are taken to have, are we ever right? Or does this always amount to prejudice and bias? As with interpreting clouds on the horizon, we can be and are often mistaken. But that doesn't change the metaphysical fact that sometimes certain features do naturally mean something.

With age we can have instances of natural meaning such as:

13) His white hair means that he is over thirty.
14) His red face means he's been drinking today.
15) Her sudden weight loss means she's been using drugs again.

We can also find natural meaning with age in the animal world, as in, if speaking of an Allen's Hummingbird:

16) His green feathers mean that he's a juvenile male.

With age in both the human and non-human animal world there are reliable indicators that we can use to gauge these things, and there might be evolutionary reasons that it was advantageous to have the metaphysical and the epistemic come together on the question of age: it is beneficial to a species to be able to tell which of its members are and are not within a child-bearing and child-rearing age.

With height the story begins to become more complicated. Height may be a fairly reliable indicator of who would win in a fight, but, as I noted earlier, it is taken to mean much more than this. It is taken as a sign that someone will or will not be a good leader, as in the following:

17) His height means he will be a commanding CEO, or
18) His height means he will not be a commanding CEO.

Could we ever in any of these cases say that there is *natural* meaning here? To say that there is would be to draw ourselves this connection between height and being commanding. Does someone's height ever make him more or less commanding?

Height and age are, of course, not the only means by which we assess the bodies of others. Race, gender, body type, disability status, and so on are all other ways by which we take the bodies of others to *mean* things. Is meaning$_N$

even possible in such circumstances? Or should we say instead that it is a conceptual impossibility, and that these are simply cases of the *misperception* of what is taken to be natural meaning? These are questions I will come to again.

The Nature of Natural Meaning

This discussion calls into question the precise nature of natural meaning. What, in fact, need be the relationship between the meaning-object and the meaning for meaning$_N$? Need there be a causal relationship? Is mere correlation enough? Some of Grice's cases do seem to be causal relations. He writes:

19)   This year's budget means we shall have a hard year.

In other words, this year's budget will cause the year to be hard because it is low. It is hard to think of another reading of this case, but we can take a stab at it. Perhaps in a past year the budget was $6,666,666 and this was taken to curse the year, and this year it is again $6,666,666 and it is again thought to be a curse. Even on this very idiosyncratic reading the relationship here between the budget and the hard year is causal, and could be restated:

20)   This year's budget will cause us to have a hard year.

But with other cases of natural meaning the relation is not causal but correlative, as in the case I have been discussing:

21)   Those clouds mean rain.

How we parse this depends on exactly what we understand by "those clouds." The clouds themselves do, of course, cause the rain, in that they are the source of the rain. But I never took this to be what this case of natural meaning was about. Rather I took it to be about some feature of the clouds, such as their darkness in the sky. There is no causal relationship between the darkness of the clouds and the rain. It is rather that the darkness is a result of their density, and when they are very dense it rains.

We see a similar situation, with another of the cases of natural meaning that Grice presents right at the start of his 1957 "Meaning" paper:

22)   Those spots mean measles.

The spots do not, of course, cause the measles but are rather a result of the virus. Thus, it would be wrong to paraphrase this case of natural meaning in the following way:

23) Those spots have caused measles.

The spots symptomatic of measles happen to be a consequence of the virus that is visible to us without medical tools. We could paraphrase this case of natural meaning as:

24) Those spots are correlated with and thus indicate the presence of measles.

Clouds meaning rain is difficult to classify as causal or correlative, and looking closer at these cases shows that natural meaning is not by any means a tidy concept.

The Distinction between Natural and Non-Natural Meaning

It is difficult to draw a sharp distinction between natural and non-natural meaning,[4] and this could be seen as a potential problem for what I argue here. The fuzziness of this cutoff point was something Grice himself recognized. He noted that in some cases we may not be sure how to classify some instance of meaning but usually we will lean one way or another (Grice 1989). However, the fact that there may be certain instances that appear to be borderline cases does not take away from the fact that natural and non-natural meaning pick out distinct kinds that prove to be useful theoretical tools. By way of analogy, to take another example, the fact that there are borderline cases between having hair and being bald does not mean that there is no such thing as having hair or being bald (Tye 1994). Some particular instances may be difficult to classify and a "commitment to sharp dividing lines" (Tye 1994) is not required. These borderline cases do not mean the categories do not exist. In short, I agree with the point about the borderline cases but do not see it as presenting a problem for my account, and Grice did not either.

As I've mentioned, Grice does not spend a lot of time detailing exactly what he means by natural meaning, and thus we are left with questions unresolved. Natural meaning acts as something to define non-natural meaning in contrast with. With non-natural meaning in bodily adornment things are more clearly and carefully presented.

3

Details on the Gricean View

Intro

The view of meaning that I have argued for here can raise some questions and lead to some objections. In this chapter I consider and respond to comments on and objections to the account I have presented thus far. In doing so I hope to also further clarify my own view. Some of these questions and objections were found in the related literature on intentionalism, some have been posed to me in presentations and in discussions with colleagues, and others still are those I give voice to and imagine a reader might have at this point. In this chapter I will also present and explain my account of imitation of natural meaning, which I see as a third category that can supplement the two categories of meaning presented by Grice.

Word Meaning and Speaker Meaning

Although I draw primarily on his work in carving out this distinction, Grice is of course not the only one to draw attention to this gap between what the words in an utterance mean, and what a speaker can mean by uttering those words on a particular occasion. This gap comes across clearly in cases of reference.

In a classic example philosopher Keith Donnellan has us consider an utterance of the phrase "Who is the man drinking a martini?" (Donnellan 1966: 287). We are to imagine that this utterance is made in a room full of people and that there is, in fact, a man drinking something out of a martini glass. Martinis are clear or sometimes cloudy liquid with a garnish of some sort. We could never really be sure that it is actually a martini in someone's glass unless we taste it or see it made. For the Donnellan case we are to imagine that it is *not* in fact a martini in the glass.

However, although no one actually fits the description of "the man drinking a martini" we all know who the speaker meant—who he referred to with this utterance—and the fact that it was a different substance in this man's glass is unimportant. If we had been watching the drink be made and knew it was a vodka gimlet, we will not reply "no one is drinking a martini" and to do so would be uncooperative and overly pedantic. The speaker was not using the phrase "the man drinking the martini" to attribute the property of drinking a martini to the man, but simply to refer to him, to pick him out of the crowd. Any competent hearer knows what the speaker meant to ask, although the man might not strictly speaking fit the description.

Along these same lines, philosopher Saul Kripke argues that this can happen not just with descriptions such as "the man with the martini" but also with names and explicitly refers to Grice in making his point (Kripke 1977: 109). In the now-canonical case that Kripke presents in the paper we are to imagine that two people see a man who they both take to be Jones, but really it is Smith. One makes the utterance "What is Jones doing?" and the other replies "Raking the leaves." Kripke argues that the two people succeed in referring to the man that they both see, although they do not use the correct name to refer to him (Kripke 1977: 111). The speaker's words "said" something false because Jones is not, in fact, raking the leaves; but, as Kripke points out, what the speaker meant by saying those words on that occasion was something true, namely that the man they could both see was raking leaves.

As the Donnellan and Kripke papers show, there is, as Kripke puts it, a distinction to be made, "following Grice, between what *the speaker's words meant*, on a given occasion, and what *he meant*, in saying these words on that occasion" (Kripke 1977: 109, italics in original). As Grice lays out in his 1957 paper "Meaning" when we interpret non-natural meaning we are attempting to work out what the speaker intends to communicate. The words he or she utters are a good indicator of that intention, but we do not limit ourselves strictly to those words. We recognize that sometimes speakers err in their utterances and seek to identify the intention beyond this. The shift that comes to light with Grice's characterization of non-natural meaning was a crucial one that was then drawn on by later philosophers of language including Donnellan and Kripke.

Grice on Cooperation and Implicature

Perhaps more well known than Grice's distinction between natural and non-natural meaning is his notion of implicature. As we see with the cases of a speaker being mistaken about what someone is drinking (Donnellan 1966) or when the person they are seeing is Smith and not Jones, that there is an important distinction between what *the speaker's words meant*, and what *he meant* by saying them (Kripke 1977).

Crucially with the Donnellan and Kripke cases, the speaker was misinformed in some way. With "the man drinking the martini" the speaker was misinformed about the contents of a glass. With "Jones" the speaker was misinformed about who he was seeing.

One of the things that makes these cases get off the ground is the fact that we can understand why a speaker would be mistaken. With an understanding on the part of the hearer that an utterer is mistaken a hearer automatically goes through the process of working out what the speaker *could* have meant or intended to refer to. We assume the speaker is acting cooperatively, realize they are is mistaken with their referring intention, and interpret them charitably to be referring to "that guy" even if he is not holding a martini or is not in fact Jones.

When we speak with others we take it as a given that we have certain communicative goals in common. This basic assumption is what Grice calls his Cooperative Principle. According to Grice the Cooperative Principle underpins all our communicative exchanges. The Cooperative Principle is that conversational partners will make their conversational contribution such as is required, at the stage at which it occurs, by the accepted purpose or direction of the talk exchange (Grice 1989: 29). This is a "quazi-contractual matter" (Grice 1989: 29). Although the speakers in the Kripke and Donnellan cases are mistaken about the facts, they are acting cooperatively and we interpret them as such; again, a hearer recognizes that they are mistaken.

Sometimes a speaker is not mistaken but instead intends to implicate something by referring to someone with a name or description that they do not actually fit. In these cases, assuming they are acting cooperatively, we can infer that they are suggesting something by doing this—this is what Grice calls conversational implicature.[1] We can imagine the following way of referring to someone's wife:

25) Your parole officer is here.

A hearer would be mistaken to simply interpret this along the lines of the "man with the martini" case and think that the speaker is mistaken and believes the wife to be his parole officer. There is something more going on here. The speaker is intentionally flouting the norm that we usually refer to someone with a description that fits them and, relying on a tired trope, thereby implicates that his wife in some way acts as a parole officer would.

We can also generate an implicature by referring to someone by the wrong name. Of course, this happens all the time unwillingly, in cooperative speakers who are simply mistaken, but sometimes it is intentional. For instance, to take a case I observed on reality television, imagine a woman named Jenny is dating a new man, Zack. Jenny's goofy friend Uncle Nino thinks Zack resembles Jenny's ex-husband Roger. They are both inordinately muscular. Let's say that when he meets Zack Uncle Nino refers to Zack as "Roger." A hearer might at first think Uncle Nino is confused but by recognizing the mischievous look in Uncle Nino's eye will understand that he was intending to implicate that the new boyfriend resembles Jenny's ex-husband. Uncle Nino was trying to embarrass Zack and Jenny. To think Uncle Nino is simply confused as a hearer is to miss out on what the speaker intended to convey with this intentional "mis-naming."[2] In this case and with the utterance of "Your parole officer is here" a hearer needs to recognize the intentional flouting on the part of the speaker. If they fail to do so the implicature will go over the head of the hearer. In summary, with implicature the speaker's intention has to "click"; there is something non-literal to "get," and the speaker is not simply mistaken as in the Donnellan and Kripke cases.

Implicature in Adornment

According to Grice, non-natural meaning requires appeal to intention and has two components: that which the utterer says, and that which the utterer *implicates*. Something is "what is said" in virtue of there being certain community-wide conventions that some word or string of words has some range of likely meanings, such as that a black armband is a protest of something. Within a cooperative system, when these conventions and intentions fail to be enough to maintain the premise that the speaker is being cooperative, following Grice's "Cooperative Principle," hearers take him to be cooperating at the level of what he implicated (Grice 1989). Implicature is not only is restricted to verbal utterances but is also found in bodily adornment.

To demonstrate, consider cases of irony. If someone says, "she's a fine friend" of someone known to have just betrayed the speaker, based on what he *said* we would take the speaker to have asserted something blatantly false (Grice 1989). However, if we maintain the assumption that the speaker is cooperating, we understand that he is merely "making as if" to say this in the course of being ironic. We can say he did mean something true at the level of what he implicated, namely, "she's not a fine friend" (Grice 1989). The conventions underpinning what-is-said are often manipulated by way of conversational implicature. If we want to say that with dress we have a similar distinction between what-is-said and the implicated content, we need to find the same phenomenon there.

The example of the foregoing implicature was one of irony; this linguistic term is very commonly used to describe clothing (Rifkin and Goodwin 2014), indicating it is already recognized that there is what may be called implicature in clothing.

One paradigmatic example of irony in clothing is seen in the phenomenon of the Ugly Sweater Party (Figure 3). Before such parties, attendees usually go to a thrift shop to find an outlandish sweater. If someone sees the dressers out of the context of the party, they might be confused because the sweater-wearer would seem to violate a number of expectations about the aims and norms of dressing.

At what we might call the level of "what-is-said" by the sweater, the dresser seems to not be cooperating. However, the dresser is wearing the sweater

Figure 3 Ugly sweaters in Amsterdam. Examples of irony in clothing worn on women in Amsterdam, the Netherlands, during a 2017 Ugly Christmas Sweater Run charity event. *Source*: Getty Images.

ironically, just as "she's a fine friend" was uttered ironically. Once we know the dresser is wearing the sweater ironically we understand that a number of the messages he seemed to have been conveying are quite different from what he means. As with "she's a fine friend" we are invited to laugh at the absurdity that someone might have meant the what-is-said content literally—to recognize that there is an amusing and stark contrast between the utterance the speaker made as if to say and how things actually are. Irony in dress is a case where we can restore our belief that the speaker is cooperating by taking them to mean other than what they at first were taken to mean by some garment.

An important assumption that underpins the implicature of the ugly sweater going through is that the wearer would not normally wear such a garment. It requires wearers and interpreters who have options about what they wear. These sweaters are a deviation from their norm.

Second-hand clothing being read as ironic relies on the assumption that the wearer doesn't ordinarily shop in these stores or wear these sorts of garments. This will not make sense to all interpreters of the garments, notably those who may need to wear such garments because of a lack of money or options. This is captured well in a scene from the novel *I Am Not Your Perfect Mexican Daughter* by Erika Sánchez. In the scene the protagonist visits a thrift store with her wealthy white boyfriend for the first time. Sánchez writes,

> After school, Connor and I meet in Uptown, at his favorite thrift store. His face is flushed from the cold, and he looks cute in his big, puffy jacket and purple stocking hat.
>
> Though I love looking at old and used things, I kind of hate thrift stores because they make me feel itchy and remind me that I have no money. For Connor it seems like a fun adventure, probably because he's never had to shop there. Amá, Olga, and I used to go to the one in our neighborhood on Mondays because it was half off. How sad is that? A sale at a freaking thrift store.
>
> "Oh my god, look at this," Connor says, and holds up an embroidered sweater with three cats on it, something an old lady would wear. "This is amazing. It's so ugly, I kind of want to buy it."
>
> I smile. "Yeah it's pretty hideous, like, disrespectful to the senses. Where would you wear that, though?"
>
> "Anywhere. I'd wear this to school, to the grocery store, to a bar mitzvah, I don't care"
>
> I have six dollars to my name, and he's gonna buy something as a joke. I know it's not his fault, but I can't help but feeling a little annoyed. I try not to show it, though, because I don't want to hurt his feelings. (Sánchez 2017: 188–9)

The narrator later says that looking through the thrift store clothes makes her think of bedbugs and she asks to leave. We see in this case a difference in the assumptions of the norms of dressing. For the narrator of the piece, she could recognize the attempt at irony but it would not have the same effect because she and Connor do not have the same background. The wearing of a sweater with three cats on it by a teenage boy will not be read as flouting in the same way by all interpreters. As with all meaning, the background of the interpreter must be taken into account when formulating an intention. Here perhaps Connor will realize that the background he assumes with his ironic sweater is not shared by all.

Under the right circumstances, and with a shared understanding of the background norms dressers are able to implicate. That is, we see some cases where hearers understand that a wearer meant other than the literal content they seemed to be expressing. These examples of irony demonstrate that systems of dress share many features of language and can be explained in great detail when considered as a part of the Gricean theory of meaning. And just as with irony in language, irony in dress can fail to be understood based on the relationship between the utterer and the interpreter.

Anything Can Mean Anything

Now that I have explained the contours of the Gricean view let me respond to some questions that may arise. My appeal to intentions as constitutively determining of meaning does not entail that anyone can mean whatever they want by whatever they do. This is true for language, and it also holds for adornment. To put it simply: there are constraints on the formation of genuine intentions.

To voice this objection with a particular case let me appeal to Lars Svendsen's 2006 book *Fashion: A Philosophy*. In the book, Svendsen rejects the idea that intentions bear on what clothing means. In a chapter devoted to fashion and language he considers the intentionalist position, writing if "meaning can be found in the consciousness of the person wearing the garment" then "a garment would then mean this or that according to what the wearer of the garment thinks it means" (Svendsen 2006: 69). Svendsen illustrates his point with the following case: "I cannot simply put on a black suit and claim that this suit says: 'I am worried about the consequences of globalization's impact on the cultural diversity of the world'" (Svendsen 2006: 69). This sort of objection might appear to be a problem for the Gricean, intentional account of meaning I have argued for here.

Of course, one can claim whatever one likes. People do lie. But likely what Svendsen wishes to illustrate with this case is not really about claiming but about intending—hence his stated target of meaning as "in the consciousness of the person wearing the garment." This sort of objection to intentionalism is not new (Searle 1969; Bach 1987; Saul 2002a, b), and Svendsen's formulation of this sort of objection is strikingly similar to a well-known case considered by Donnellan (Donnellan 1968).

In a canonical paper about intentions in language (Donnellan 1968), Donnellan considers a passage by author Lewis Carroll in which Humpty Dumpty utters "There's old glory for you" and says that he thereby means "There's a knockdown argument for you" (Donnellan 1968: 211–13). Of course, Humpty Dumpty cannot get away with this. Meaning does not work in this way.

As Donnellan goes on to argue, the Humpty Dumpty case does not pose a problem for a theorist who places intentions at the center of a theory of meaning. Humpty Dumpty is a competent speaker of the English language. He could therefore not formulate a genuine intention to mean "There's a knockdown argument for you" by uttering, "There's old glory for you." That is, because Humpty knows that there is no way his audience will take him to have meant "There's a knockdown argument for you" by his utterance of "There's old glory for you" he could not have genuinely meant it. As Donnellan concludes in the paper, there are constraints on the sorts of genuine intentions one can form with respect to some utterance, some meaning, and some audience (Donnellan 1968: 211–13). We cannot go around intending to mean whatever we want. This is true for words and for items of bodily adornment as well.

We can make *claims* about what we meant, but this is disingenuous, and is not the same as having a genuine intention. Svendsen is right that without some sort of elaborate setup one cannot have a genuine intention to mean that one is concerned about globalization by wearing a suit. But this fact does not pose a problem for an intentionalist.

In other types of circumstances, the stage can be set such that we *are* able to form a genuine intention to mean something quite specific and unexpected. In contrast to the Svendsen suit case, with the *Fawlty Towers* black armband case the wearer would have had the reasonable expectation that the audience would work out what he meant by the armband. The connection between garment and meaning in the *Fawlty Towers* case is no less improbable than in the Svendsen case—but the circumstances were such that the speaker was able to form a genuine meaning-intention. The audience had to have a great deal of knowledge to correctly work out what was meant by the armband, and we, as viewers of the scene, were able to correctly interpret it. However, without this careful setup by the writers the plot could not have been brought to this point. Meanings are not easy to achieve.

Mistaken about Meaning

Sometimes we as speakers or wearers aren't being disingenuous about meaning, but are simply mistaken. With clothing, as with words, we sometimes can say that a speaker or wearer is acting or speaking erroneously. If one is misguided about what some items of dress can mean, such as in thinking that punk clothing will signify elegance at the Metropolitan Opera,[3] we will recognize their error and perhaps correct them.

If someone thought the word "punk" meant spiky and said "This pineapple is punk" we would say he was mistaken about the meaning of the word and what sorts of things he could reasonably expect hearers to take him to mean by such an utterance. However, he did mean "This pineapple is spiky," even though he uttered the wrong word to get this across. Once his error is pointed out, he will no longer be able to develop this intention in the future. We can certainly be misguided in our intentions, or fail to achieve our aims. We should attribute this error to the speaker's confusion and not to a problem with the explanation of how intentions are constrained. A cooperative speaker who is corrected will subsequently change their use of the word, or could perhaps again flout the meaning to make a self-deprecating joke to a knowing audience.

Nonlinguistic Communication

I have argued here for the benefits of understanding communication through bodily adornment in terms of the Gricean framework. An objection may arise to my application of the Gricean framework to nonlinguistic communication. Although there is a wide range of communicative acts that can be performed through bodily adornment it is clearly not a form of *linguistic* communication. Philosophy of language could potentially be understood as a project of explaining purely *linguistic* phenomena. In proposing to draw on H. P. Grice's philosophy of language, as I do here, we need to consider whether the relevant machinery can be carried over intact to nonlinguistic phenomena.

There would be a flaw in the premise of such an argument: it is not evident that there is a non-arbitrary line between the linguistic and nonlinguistic to be drawn. This is an issue that applies to not only human communication but animal communication as well. Let us consider this point historically: in the 1970s Karl von Frisch won the Nobel Prize for his work on "the language of the bee." Bees communicate where to fly to find nectar with their bodies, doing "dances" (von Frisch 1974). The dances are a complex form of communication whereby the bees share the angle and distance to food with stunning accuracy (von Frisch 1974). Perhaps we should follow von Frisch's characterization and call this behavior language. Or, one might say this cannot be the case and unreflectively suggest that languages are restricted to vocalizations. But, of course, this would rule out sign languages—an absurd consequence (Senghas 2004).

Even with the language that we take to be uncontroversial, it becomes clear that there is some fuzziness in however we try to draw this line between the linguistic and the nonlinguistic. This difficulty is especially apparent with certain words called demonstratives: words such as "that," "this," and "here." These words rely on the interaction between the word itself and something else—it could be the linguistic device of anaphora—referring to something previously uttered—or it could be a nonlinguistic physical gesture on the part of the speaker.

Imagine someone performing the act of requesting a certain baked good at a bakery while they utter, "I'll have that." Most of us lack the full vocabulary needed to individuate one type of bread from another and must resort here to gesture. In the use of the linguistic term "that" in conjunction with the act of pointing the linguistic and the nonlinguistic each play a vital part in a speaker referring. The speaker must make her intentions manifest to the hearer in some way (Kaplan 1989). With indexicals such as "that" intentions are sometimes made clear not with pointing but with the hearer tracking the speaker's gaze (Kaplan 1989; Baron-Cohen 1995; Tomasello 2010), clearly a nonlinguistic phenomenon. These cases present a problem for any theorist who purports to work only with the linguistic. Aside from artificial languages such as computer code, the "pure" linguistic that doesn't rely in some way on the nonlinguistic is a bit of an illusion.

Fortunately for this project, the work of Grice, in particular, is not restricted to the realm of the linguistic, and so drawing on Grice's theory in discussing adornment is a natural fit.[4] Grice holds the view that "the principles at work in the interpretation of linguistic behavior are (or are intimately related to) those at work in interpreting intentional *non*linguistic behavior" (Neale 1992: 20). Because of this connection Grice sees between interpretation and speaker meaning we can take this point to extend not just to the interpretation of meaning, but to the creation of meaning. So, we can say that the principles at work in the *formulation* of meaning-intentions with respect to linguistic utterances are (or are intimately related to) those at work in the formation of meaning-intentions with respect to *non*linguistic utterances.

We can see through this discussion that the theory I draw on here can be appropriately applied to nonlinguistic meaning. But aside from what Grice thought, more importantly, I hope to have provided general support for the notion that linguistic and nonlinguistic meanings are both importantly grounded in intentions and *ought* to be considered as opposite sides of the same coin.

Personal Stylists

Consideration of the nuances of dressers' intentions might lead us to wonder about cases where a dresser uses a personal stylist. Can a wearer mean something if they did not choose the garments themselves?

Again, this case draws into focus the close connection between adornment and utterances. In the case of a personal stylist, we see a parallel with having a speech writer. We could ask, "Can a speaker mean something if he/she did not write the words themselves?" With a stylist, as with a speech writer, the wearer or utterer is in a position of power. They employ the stylist or speech writer because the stylist or speech writer possesses a special skill they lack, or perhaps don't have time for. In both cases, although someone else chose the actual garments or words, the person uttering them or wearing them has a choice about whether or not to wear or say what they are given. They also share communicative goals with the writer or stylist.

A person using a stylist, be it a celebrity or a shopper at Macy's, has the power to say "no" to any garment they are presented with. Sometimes there is a gap between what the person says they want to communicate and what the stylist thinks will best do this, or with what the person is comfortable wearing. The two may have a disagreement and the stylist may think the client is making a mistake. The client may later on realize the stylist was right.

Again, we see the same thing with speechwriters. To take a specific example to illustrate the point, Bill Clinton's speechwriter thought he should have come clean in his infamous 1998 "I did not have sexual relations with that woman" speech, but Clinton chose to override the prepared remarks (Glass 2020). Clinton ultimately was responsible for the speech he ended up giving; blaming it on the speechwriter would never be convincing, because we know Clinton had the final say. The same goes for celebrity stylists (Wallace 2018). Just as speeches can go wrong, as Clinton's did, so can ensembles. But in either case we know this person chose to hire that stylist or speechwriter, and did not veto some part of the ensemble or speech. This means that the final product is ultimately the product of their intention.

Surely there are many lingering issues of intentions in such cases to be further explored that could illuminate what might be called "The Problem of Multiple Authorship" for communication through dress and through language. This is also an issue for intentionalism in film and art. I will not address these questions further here but highlight the fact that this is essentially the same question arising for dress and language. This seems to provide further evidence for my main thesis—that adornment and language are best understood in tandem.

Uniforms

With uniforms, unlike with personal stylists, quashing the personal expression of the wearer is not just a side effect but a goal. This is why they can be so demoralizing or empowering to the wearer, depending on the uniform. I have personally worn both demoralizing and empowering uniforms. My favorite was when I was a basketball referee in High School and in putting on that striped shirt I had a power I usually lacked in my everyday life. In this uniform I could tell a grown man to stop yelling and sit down or he would get a second technical foul. I distinctly remember the power I felt ejecting the few people from games who had crossed this line. With this striped shirt I was no longer "just" a teenage girl, but "the ref"—someone with power over everyone in the room, and my uniform made this clear.

Similarly, I also remember wearing uniforms that I found demoralizing and "not me" in a bad way. As a waitress at a Yacht Club in High School I had to wear a white shirt with a tie, a black skirt, nylons (yes, it was in this century), and black shoes. I was told by the manager that these were "better than the last uniforms," which apparently had ruffles and made you "look like a French maid." Again, in the uniform I wasn't an individual, but "the waitress." On occasion I would be mistaken by the wrong table for their waitress; we were seen as interchangeable.

It is no accident that in a uniform I felt like "not me," either with more power or less. That is the point of a uniform. The wearer of a uniform is not acting in the world as an individual; this is not relevant. Now, as a professor, I wear what I like, within certain limits. But many workers spend the majority of their working lives constrained by uniforms.

"Slaves" to Fashion

In their work, philosophers Lauren Ashwell and Rae Langton (2011) have argued that the ways fashion shapes our beliefs about what looks good is a form of pernicious constraint on wearers. On their account, we are subject to certain "aesthetic restrictions" that dictate which garments and compositions are pleasing to us (Ashwell and Langton: 2011). Those garments that are pleasing we deem to be fashionable, without recognizing the ways that this judgment was outside of our control. This is what the authors call a "projective illusion." Ashwell and Langton use this line of reasoning to conclude that we are all in some sense "slaves to fashion," which is the title of their piece.

If we understand clothing as a part of philosophy of language, as I propose here, there is no reason to think that these constraints on what wearers think looks good or what they can communicate through dress presents a unique problem. These restrictions are better understood as analogous to those that prevent the development of the intention to communicate messages that the speaker knows the hearer is unlikely to understand. For instance, we would say that Humpty Dumpty is constrained in the sense that he cannot form the genuine intention to mean "there's a nice knockdown argument for you" by uttering, "there's old glory for you" (Donnellan 1968). However, we would not consider this to be unfair to Humpty Dumpty or say that he is "enslaved" by the meanings of the words he utters. These constraints are simply a product of the way conventions of communication work—the same conventions that help Humpty Dumpty to get his point across when he utters the right words.

At the same time, this is not to say that fashion presents a level playing field or to naively assume that the experience of wearing adornment and being interpreted is the same for everyone. People have different bodies and this affects how others interpret what they are wearing. It may be nearly impossible for people with certain bodies to be taken seriously, or for people with other bodies to be viewed as non-threatening. Some may not give a second thought to wearing a type of garment—such as a facemask, or a black hoodie—where others will recognize that this could increase their chances of being perceived as a threat (Jeffers 2012; Taylor 2020).

During the coronavirus pandemic it became commonplace to wear a facemask (and as of writing this it still is). We saw the conventions surrounding this garment change. In an article on race and masks at the start of the pandemic it was noted, "African-American men worry that following the

C.D.C. recommendation to cover their faces in public could expose them to harassment from the police" (Taylor 2020). Indeed, this worry is clearly justified because, as documented in the article, in March of 2020 two African American men in Illinois were kicked out of a Walmart by a police officer for wearing masks.

This is not just a matter of an "aesthetic restriction" about what looks good but a matter of safety. Women, too, may ask themselves if certain garments are physically safe to wear in certain situations. In Chapter 6 I will return to this issue and will consider how what we are taken to communicate with dress and our bodies is unequal in ways that are outside our control. I will also describe attempts to take back some of this power—a process I frame within work on metalinguistic negotiation. The issues that arise through consideration of the point made in the Ashwell and Langton piece are important and not resolved simply.

Adornment Is "Mere" Imitation

A related, more ancient objection to our practices of adornment, construed broadly, comes from Plato, who saw the subject as worth commenting on. Plato argues that clothing is merely imitative (Pappas 2008; Pappas 2015: 8). In work describing Plato's views, philosopher Nick Pappas writes that fashion was "understood as everyone's attempt to resemble everyone else" and this was seen as resulting in "pathological attentiveness to one's fellow humans" (Pappas 2015: 8). The worry that dress is governed by rules of "mutual mass imitation" (Pappas 2008; Pappas 2015: 8) appears to carry some weight when dress is considered within Plato's framework of the Forms and as a sort of art.

However, Plato's point becomes weaker when we consider dress to be a part of language. What had been seen, perhaps reasonably, as "copying" along the lines of a "slave to fashion" instead becomes the repetition needed to establish conventions and meaning. Is language a case of "mutual mass imitation"? Perhaps we could call it that—but with language it seems freed of its negative connotations, and if dress is understood as a part of language then it, too, is freed. We can say instead that many people wear more or less the same clothes because for each new deviation, enough observers must understand the new meaning for the wearer to formulate a genuine intention, and this is a slow process with dress, as it is with language.

Conventions

Given my argument thus far, an objector might ask, "Isn't this really a matter of conventions, not intentions?"[5] In the previous chapter I wrote that the wearing of a black armband in mourning originated in the practice of wearing the color black for mourning more broadly. In Victorian times, widows—including Queen Victoria herself—would wear all black clothing after the death of their spouse (Hollander 1993; Death Becomes Her). Does it make sense, then, to say that a black armband means what it does in virtue of these conventions, rather than the wearer's intentions?

We see with items of adornment, as with words, meaning changes and evolves over time. With items of adornment, as with language, we might at first be surprised to learn of the original meaning, but then see the connection upon further reflection. To take an example from language, the French word *pas*, which now indicates negation, took its original meaning from what is now a secondary meaning. *Pas* also means "step"—as in the ballet term for a duet *pas de deux*—and was originally used for emphasis in sentences such as "*il ne marche pas*"—"he doesn't walk a step" (McWhorter 2003: 26). The term *ne* here was the original negation, playing the same role as our similar-sounding words "not" and "no." The French *il ne marche pas* is similar to English phrases such as "Do not move a muscle" or "Do not move an inch." One could simply state "Do not move" but with the addition of "a muscle" or "an inch" emphasis is added. In the French construction eventually this emphasis that had been particularized to certain utterances expanded to all negation. Now the *ne ... pas* construction is standard "proper" French for negation (McWhorter 2003: 26-7).

This means that there is a certain redundancy in the language because the negation is indicated twice. And when there is redundancy in language we tend to shorten things for the sake of efficiency (McWhorter 2003). This happened in French. As I was initially surprised to learn upon moving to France, in much spoken French the *ne*, which I had learned in school to say, is dropped all together as in *J'ai pas le temps* for "I don't have time." The word that originally meant "step" remains as the marker of negation. This history is interesting, but it does not tell us what the term *pas* means today; the meaning of words changes over time.

The meaning of items of adornment changes with time as well. The garment with perhaps the longest history of relatively stable meaning is the Muslim veil, or "hijab." This item of adornment has a similarly long and rich history to the French word *pas*. Muslim women began to wear headscarves over a thousand years ago, in the ninth century (Hajjaji-Jarrah 2003). At the time, slave ownership was commonplace, and female slaves were often sex slaves (Hajjaji-Jarrah 2003). Because of this, Al-Tabari, a commentator on the Qur'an, saw a need to distinguish Muslim women from the sex slaves, who were not Muslim, and it was the veil that played this role (Hajjaji-Jarrah 2003). This means that, at the time, the veil could be seen as communicating that the wearer is not a slave.

Of course, the meaning of the headscarf, as with the French word "pas," can change meaning over centuries of use. Anthropologist Homa Hoodfar has studied what Muslim women in Canada take themselves to be communicating by wearing the hijab today. One of her subjects commented, "this scarf, that to so many appears such a big deal, at least has made others aware of Islam, and of my identity within the Canadian society, instead of looking at me and judging me for my figure and looks" (Hoodfar 2003: 30). To this woman, the scarf serves to obscure the body and announce her religion in the Western country of Canada. Hoodfar comments that "by taking up the veil" these modern Canadian woman "symbolically but clearly announce to their parents and their community that, despite their unconventional involvement with non-Muslims, they retain their Islamic mores and values" (Hoodfar 2003: 21). These contemporary meanings in modern-day Canada are quite removed from the historical setting in which it was first determined that Muslim women should wear such coverings, but the connection is not fully severed.

The Muslim veil shows the contrast between what Muslim women in Canada today take themselves to be communicating by wearing the hijab, and original intended meaning of the garment when the practice began a thousand years ago. This should not be surprising given the many parallels we have seen between adornment and language. The meaning of adornments like the hijab should be expected to change over the course of a thousand years, just as the meanings of words like *pas* do.

There is this history, which perhaps establishes what we want to call a "convention," but simple appeal or reference to this history does not tell us what the garment means today. Further, conventions themselves are not enough to individuate particular meaning on some occasion. With a black armband the convention that it is usually worn in mourning is not enough for a viewer to determine what it is that is being mourned or protested. The same was the case for the bulldogs jersey that I discussed in a previous chapter.

For one more illustration of this point, consider uniforms again. The uniform for workers at Target stores is a red shirt and khaki pants (Johnson 2019). This is the conventionalized meaning of a red shirt and khaki pants because the workers wear this combination when they are working at the store. But perhaps you have had the experience of walking into a Target and realizing that *you* were dressed like a Target worker. What a surprise to suddenly find yourself in an environment where what you are wearing is taken to have a meaning you did not intend. Your clothing does not mean you are a Target worker if it was an accident that you went to the store in red and khaki. This is despite the fact that you are wearing something with this meaning "convention" in the right context. Such cases further illustrate the point that conventions are not enough to individual meaning on some particular occasion. Conventions do play a role, but they do not determine meaning. Conventions are something that shapes the formulation of a wearer's intentions—which is ultimately what meaning depends on.

Saying Nothing

One might insist that they do not mean anything with what they wear. Yes, we are obligated every day to make a decision about how to dress, but some may attempt to "opt out" of the communicative system of dress by wearing something "neutral." Depending on where you live, perhaps this means an ensemble of all black.

However, such attempts do not truly fail to communicate. A uniform of all black, popular with certain art directors and philosophers alike, inevitably conveys a sort of cosmopolitan ennui—albeit one so common in certain circles that it goes unnoticed. One need only wear these "normal," "neutral" clothes in a new environment—such as Miami Beach, or Southern California—and the wearer will quickly realize that what was "neutral" or "normal" in one context fails to be in another.

Again, we find a similar story with language. We cannot always "opt out," or "say nothing," even with silence. If someone is insulting our friend we have an obligation to defend him or her, and to stay silent can rightly be seen as a betrayal. Similarly, as philosopher Liz Camp has highlighted in her work, in not objecting to the use of a slur, we may be complicit in supporting the attitudes expressed (Camp 2013; see also Anderson and Lepore 2013). We should consider bodily adornment to be a similar case where communicating nothing is not an option.

"I don't think about what I wear"

An objector might push back against this point and insist that for them, they truly are neutral and they do not think about or mean anything by what they wear. This has been posed to me when presenting this material in person. Once, when I gave a presentation in Australia on this material I got this question during the Q&A from a man who said he doesn't think about what he wears. He said he just always wears black shoes, black pants, and a band T-shirt.[6]

What are we to say of meaning in this case? I replied to the man that it sounded like in fact he had a very particular look (he, of course, had it on then), and I bet there is a reason he wears black pants, black shoes, and a band T-shirt. I bet he also was very particular about which band shirt he wore, and perhaps he had gotten that shirt at a concert, creating a connection between the garment and a happy memory. It seems unlikely he would wear a band T-shirt from a band he hated or found embarrassing, such as one he got as a gift from his mother who tries but fails to understand his taste in music. He agreed that this was the case. Just as traveling to a new place can reveal that our "neutral" clothing is no longer neutral, sometimes gifted or loaned items of adornment can reveal the extent to which we do care about what we wear.

I was again asked a version of this question after another presentation of this material, this time at a conference in Croatia.[7] I was asked it by a man wearing essentially the photo negative of the man in Australia: khaki pants, a white shirt, and a white hat with red stitching outlining a symbol and some writing, which was in Croatian. The person wearing this ensemble was not native to Croatia, and I asked him to tell me about his hat. He explained that he was borrowing it from someone else at the conference because the previous day his head had gotten sunburned under the Mediterranean sun. I asked him if he knew what it said, and if he had inquired about this. And indeed he had. I can't remember exactly what the symbol was or what the lettering said but it was something benign. Of course, in this situation we would all like to be sure that we are not unwittingly wearing the Croatian equivalent of a MAGA hat, or something with an otherwise stupid message such as "YOLO" or "Ball State Alum."

Think, too, of the ease with which a woman might borrow a man's clothing in a pinch—and in fact certain baggy women's clothes are marketed as a "boyfriend" cut to give the appearance of having been borrowed from a man. In contrast, a man might be very worried indeed about the effect if it seems he needs to borrow a piece of clothing, such as a sweater, from a woman. Most will not attempt it—even those who protest most loudly that they do not care what they wear. In fact, such people may be the least likely to wear gifts because their professed "not caring" actually amounts to having a quite rigid look from which they do not deviate. To deviate for some might feel like it shows a certain frivolousness that violates ideas about the extent to which one is allowed to focus on physical appearances. For some the feeling may be so strong that it feels morally objectionable to deviate from their look.[8]

Artifice

Still, an objector might insist that he takes no part in fashion whatsoever, and just wears normal clothes with no message. Ralph Waldo Emerson proclaimed that for him "No fashion is the best fashion" (Hanson 1998: 156). Lois Banner, in her 1984 *American Beauty: A Social History,* calls for a rejection of "artifice" and a return to the "natural" female state of "healthy bodies and useful lives" (Banner 1984; Steele 1984). However, this idea that one can wear something "natural" or that is a case of "no fashion" is completely illusory. In reviewing Banner, fashion historian Valerie Steele writes that, "the 'natural' human body does not exist, but most people accept as natural what they are accustomed to seeing or what they would like to see" (Steele 1984: 301). There is no such thing as adornment that escapes being a part of a system of meaning just as there is no such thing as a body that is more natural than any other. The fact that there may seem to be results from overestimating the degree to which our own choices are the "natural" or "normal" ones.

The idea that there is such thing as "natural" or "no fashion" clothing is an illusion brought about by being unreflective about the ways our choices are shaped by surrounding forces—akin to holding the position that one "doesn't have an accent." To bring the point into focus we only need travel to a new place, or reflect back on "regular" garments from other points in history, which to us appear "unnatural," or maybe even absurd.

Further, say the "natural" body really *were* an ideal. Let us for the moment understand the most "natural" body to be the one that has the least adornment and grooming, allowed to return to its "natural" state and consider what such a body without artifice would be like. Think about the time you went the longest without showering, brushing your teeth, shaving, getting a haircut, changing your clothes, and so on. This was you at your most natural. Is that something to aspire to?

Perhaps it is for this reason that poet and aesthete Charles Baudelaire has declared that adornments should not be shunned because they are a form of artifice, but celebrated because they "redeem the body from its natural state" (Baudelaire 1863/1993; Hanson 1998: 158). Adornment for Baudelaire is "like the amusing, teasing, appetite-whetting coating of the divine cake" without which we are merely a body—"indigestible, tasteless, unadapted and inappropriate to human nature" (Baudelaire 1863/1993: 392). Beauty is far from trivial in the estimation of Baudelaire. Indeed, on this subject, Baudelaire quotes nineteenth-century author and dandy Stendhal, who said, *"le Beau n'est que la promesse du bonheur"*—the beautiful is nothing but the promise of happiness[9] (Baudelaire 1863/1993: 393; Gilman 1939: 295).

4

Deception in the Human and Animal Worlds: Imitation of Natural Meaning and Lying with Non-Natural Meaning

Introduction

With communication by our bodies and the way we adorn our bodies comes the possibility of lying and deception with our bodies. This applies to both natural and non-natural meaning. With non-natural meaning, the mechanisms by which we do this are very similar to ordinary language. If I am a con artist trying to get others believe I went to Yale Law, walking around with a Yale Law baseball hat will be a good start. If I would like to get others to believe I am a surgeon, I could buy some green scrubs and wear them to lunch. For this to succeed, I must hide the fact that I am attempting to deceive the viewer.

Lying or deceiving with non-natural meaning parallels how we lie with language. Lying or deceiving with natural meaning is a bit more complicated, and parallels again the animal world. When the fur of bears stands on end when they are in a fight this creates the visual illusion that they are larger and thus more physically imposing. I call such manipulation of bodies—and thus the associated meanings we have with bodies—imitation of natural meaning. Many of the ways we dress change how our bodies look and thus contribute to imitation of natural meaning. The suit has persisted for over 200 years because it appears to broaden men's shoulders, diminish their stomach, and lengthen their legs. This leads to a physical form that we have positive associations with. I present a number of instances of imitation of natural meaning at the end of the chapter.

A Compulsive Con Man

Let me begin with a case of con. On January 4, 2016, Jeremy Wilson, a compulsive con artist, walked into a New York City police station to inquire about a compounded BMW wearing "a Harvard sweatshirt, a 'Wounded Warrior' cap, and military dog tags" (McKinley Jr. and Rojas 2016). These items of dress corresponded to the persona he had recently adopted, as Jeremiah Asimov-Beckingham, a Harvard alumnus and military veteran (McKinley Jr. and Rojas 2016). Both of these aspects of his adopted identity are unambiguously conveyed by the bodily adornment he wore to the police department.

Upon his eventual arrest the items seized by the police included two Purple Hearts, two Bronze Stars, American and Canadian military fatigues, two Harvard University hats, one Harvard Law hat, one Harvard military hat, one MIT hat, one army hat, one Canadian military hat, one army T-shirt, two Purple-Heart veteran license plates, two passports, four hard drives, and so on (McKinley Jr. and Rojas 2016; Figure 4). Surely the reason these items were seized is that they

Figure 4 Items seized from Jeremy Wilson's apartment. We see in this image clothing and other forms of adornment that were taken as evidence of Jeremy Wilson's behavior as a conman. Seized were hats, shirts, passports, money, hard drives, mugs, and military badges. On the license plates we see an image of Purple Hearts, as well as Purple Hearts themselves. *Source*: Photo courtesy of Homeland Security Investigations New York.

are taken as strong evidence that Wilson had adopted the persona of someone who had accumulated these possessions through life experiences. Although there is no rule against, say, wearing a Harvard Law sweatshirt if one has not gone to Harvard Law, someone wearing a Harvard Law sweatshirt will be taken to mean that they went to Harvard Law. Because Wilson was a con artist and did not go to Harvard Law, we can take this to be a case where the speaker intended to convey false content to the audience through dress.

Lying and Deception through Dress

Contrasted with irony in dress—as discussed in Chapter 3—are cases where the speaker does not intend the hearer to recognize that the what-is-said content cannot be truthful: cases of lying and deception through dress. We even have a special word for it: disguise. Someone lies when they attempt to get the hearer to believe something false by hiding the fact that they are not abiding by the Cooperative Principle. Someone is wearing a disguise when they use dress to express content that is not true of them and are not implicating something true. Bank robbers in a Hollywood movie might wear the uniform of the janitors for a heist. This is because they want to communicate that they are janitors and that there is nothing unusual about them being near the safe. Another example of lying through dress would be, say, a military commander who wears a pin he has not won on his jacket, with the intention to communicate to others that he has. In these cases we have a determinate content that the speaker knows to be false and has the intention to convey to the audience through dress but he does not intend the hearer to recognize it as false.

So, we can say that dressers are able both to implicate and to lie[1] that is, to lead the hearers to understand that they meant other than the literal content they seemed to be saying. These examples of irony and lying demonstrate that systems of dress share many features of language and can be explained in great detail when considered as a part of the Gricean theory of meaning.

This can also happen with natural meaning. The intention is not relevant to natural meaning in bodily adornment—but sometimes people try to deceive by intentionally hiding their emotional state.[2] Being nervous is almost never a quality that we wish to convey. But confidence is not so easy to fake; there are telltale signs. This is why we take steps to make it so that our sweat is not visible at a job interview or during a presentation. This can be as simple as wearing antiperspirant or can extend to wearing certain colors or types of fabric. Some may even have Botox injected into their underarms to stop perspiration, as model Chrissy Teigen recently filmed for her 25.3 million Instagram followers, writing, "Truly best move I have ever made." In this way we sometimes attempt to hide things that would be taken to have natural meaning. Other times we *add* things that are taken to have natural meaning, such as blush.[3] And sometimes, as with being sweaty after a workout, our bodies are reliable indicators of some state of the world, of who we are in that moment, of what we've done.

Imitation of Natural Meaning

In many cases, bodily adornments make it look as though the wearer has certain natural features such as tallness and thereby alter what people perceive to be the aspects of natural meaning that are associated with those natural features: this is what I call "imitation of natural meaning." Distinct from true natural meaning in clothing, imitation of natural meaning is the way adornment makes it look as though the body has different natural features—with different associated natural meaning—than it in fact has.

Examples of bearers of imitation of natural meaning are black that makes one look slimmer and white that makes one look wider, shoulder pads, high heels, toupees, shoe lifts, hair dye, tanning spray, concealer, blush, contouring, and so on (Figures 5–6). In each case the item gives the illusion that the wearer possesses some natural features.[4] Imitation of natural meaning results from the way different shapes change how the human form plays on our eye and the various associations we have with that appearance. Imitation of natural meaning need not be factive, and in this way is distinct from natural meaning.[5]

Imitation of natural meaning does not have its effects in virtue of the viewer recognizing the intention of the wearer. Indeed, if best carried off, the intention is completely concealed and the garment will not be recognized for its imitation of natural meaning. Along the same lines, intention of the wearer is not a necessary precursor to imitation of natural meaning. Although the word "imitation" may seem to suggest intention, intention on the part of the wearer has no effect on whether the item does or does not imitate the natural meaning: One with a sore throat may or may not intend to communicate to their audience that they have a cold when speaking—their raspy voice naturally means this regardless. That some garment imitates natural meaning may even be discovered by the dresser when he, say, notices people ask if he is ill when he wears his favorite yellow

Figure 5 Hailee Steinfeld in Prabal Gurung at the Met Costume Gala, 2014. A close look will show that the wearer's form is not, as it first appears, delineated by the white fabric of the dress, but by the black panels on the side, which fade into the surroundings. *Source*: Getty Images.

turtleneck. This means that, in contrast with Barthes's attempts to systematize non-natural meaning, we expect that with imitation of natural meaning there *is* systematicity in how certain shapes and colors—for example, horizontal stripes, yellow coloring, black side panels—change the way the body appears *across contexts*. This is an important point of contrast between imitation of natural meaning and non-natural meaning.

Positive imitation of natural meaning can also help explain the persistence of certain garments. If there are advantages to appearing taller and slimmer, then items such as high heels and belts at the waist—that confer what is considered to be positive imitation of natural meaning—will persist.

Perhaps the most enduring instance of this is the suit, a "uniform so fundamentally successful that its basic elements have endured unaltered for

Figure 6 RuPaul at the MTV Video Music Awards, 1993. Contouring with makeup is seen along the hairline, nose, jawline, and cheekbones. *Source*: Getty Images.

centuries" (Trebay 2014). Suits have been a common article of dress for men in the West since their development by English tailors around about 1810 (Hollander 1994: 89; Stafford 2005: 139). Prior to the modern suit, the dress of a gentleman cut quite a different line on the wearer (McNeil 2018). Fashion theorist Anne Hollander writes,

> From 1650 to 1780 men's shoulders ideally looked very narrow and sloping and their chests somewhat sunken, and that even on slim figures the stomach swelled out prominently between the open coat-fronts and above the low waist of the breeches. This dome-like shape for the mid-section was emphasized by the descending row of waistcoat buttons that marched down its center, echoed by the coat buttons and buttonholes on either side. . . . The entire effect tended to emphasize a man's hips, belly, and thighs, shrink his chest and shoulders, lengthen his torso and shorten his legs. (Hollander 1994: 83)

The period of English fashion ranging from approximately 1760 to 1790 is known as the "macaroni" period (McNeil 2018). The line of the American song "Yankee

Doodle," "Stuck a feather in his hat and called it macaroni" is a disparaging reference to this style (McNeil 2018: 13). It is at the end of this period that we saw a remarkable shift in men's fashion away from breeches and prominent waistcoats toward suits. In his book *Pretty Gentlemen,* historian Peter McNeil writes,

> The "macaroni period" allows us to study more closely the swiftness of these changes in the wardrobes of men, as they shifted from wearing tightly cut and short French-style suits to long, informal "frock" coats with Anglophile accessories and soft hats. Their whole bodily appearance changed, from the rather pear-shaped man of the middle third of the century, with his courtly accessories of slipper-like shoes, snuff-box and cane, to the tall, lanky elegant of the end of the century, wearing long boots to the knee, droopy riding hat and carrying a riding crop. (McNeil 2018: 11)

What could have caused such a drastic change in men's fashion?

According to Hollander, the advent of the modern male suit and the imitation of natural meaning it confers corresponded with the rise of Neo-classicism (Hollander 1994: 84–90). With the burgeoning interest in ancient Greece and especially the arrival of the Elgin Marbles to London in 1806 (Figure 7) the public had access to a new model of the ideal male form (Hollander 1994: 85). Thus, the beginnings of the modern suit were cut so to evoke a "male body that tapers from broad shoulders and a muscular chest, has a flat stomach and a small waist, lean flanks and long legs" (Hollander 1994: 84). According to Hollander, the lines of a modern suit—the lengthening provided by the solid color of the legs, the widening provided by the collar and structured shoulders—have the result that the wearer of a suit appears to have a form more like the heroic male nudes of classical antiquity (Hollander 1994: 84–90). In a discussion of Hollander's work, philosopher Martha Nussbaum writes, "the eroticism of the suited body is no mere coincidence: it is created by the tailoring itself" (Nussbaum 2012: 143). In other words, this means that the staying power of this suit can be explained in terms of its successful imitation of natural meaning, a need that arose out of new ideals of the natural male form. Differences in the ideals of the natural form can explain changes in dress that evoke this new ideal.

This emphasis on the erotic nature of men's clothing in the nineteenth century by Hollander and Nussbaum is in contrast to other characterizations of men's suits found in the literature. In her work, historian and philosopher Evelleen Richards, who I draw on a great deal in her telling of this history of sexual selection in subsequent chapters, sees men's clothing of the period as "de-eroticized" (Richards 2017: 224). In discussing gender dimorphism in how

Figure 7 *Reclining Dionysos* from the Parthenon East Pediment, Elgin Marbles. According to Hollander, it was the positive imitation of natural meaning of this sort of male form that has led to the enduring success of the modern suit. *Source*: British Museum.

bodily adornment is used to signal taste and sexuality she writes, "In the course of the nineteenth century, fashion was thoroughly feminized. Men's clothing became duller and uniform, while women's clothing became more ornate, colorful, and expansive" (Richards 2017: 200). She notes that men at the time—including Darwin—have seen fashion as frivolous and feminine. Of course, such notions persist today, and are seen in some women as well, as seen in my discussion of the "vices" of artifice in the previous chapter. We will see instances of this in later chapters as well.

It is undeniable that many men today and throughout history have viewed fashion as something they ought not associate themselves with. For heterosexual men there often is a fear of not appearing masculine enough (McNeil 2018) and if fashion is thought of as gendered then to reject fashion is to be a "real" man. It also reflects the privilege many men in power have to not have to conform to others and can be seen as a flouting that demonstrates power. Think of Steve Jobs and Mark Zuckerberg here in their casual clothes. *New York Times* chief fashion critic Vanessa Friedman explains that there is a well-established tradition in Silicon Valley of tech entrepreneurs acting as if they could not care less about what they wear, including the hoodie-and-Adidas-sandal-sporting Mark Zuckerberg of Facebook and Dennis Crowley of Foursquare, who once attended a black-tie dinner in a zip-up sweatshirt and dirty sneakers. The obvious point is that said baby geniuses are too busy thinking great and disruptive thoughts, and coding through the night, to spare a moment to worry about as mundane an issue as image. (Friedman 2014)

However, even someone as powerful as Zuckerberg isn't immune from having to wear a suit in all contexts. In what may be seen as a sign of submission Zuckerberg caused a stir when he dressed up and wore what was labeled an "I'm Sorry" suit to congress in 2018, when Facebook was under investigation for "failing to protect user's data" (Friedman 2018). Friedman declared that it was "a growing-up moment, because in the iconography of clothing, the suit is the costume of the grown-up" (Friedman 2018). Given these factors, we should not take men who consider themselves to be above fashion at their word. Even the most powerful show their hand on occasion. In many situations, there is an incentive for men to underplay how actively they engage in or think about fashion.

As much as some may like to deny it men are not outside the realm of those who are judged on the basis on their bodies. A man who tries to move up in the corporate world, for example, has to spend a great deal of time and money on what to wear. To fail to do this will result in one being left out of the club. As business professor Alison Taylor notes, "It's such a fraught, unspoken signifier of whether you do or don't fit in" (Nguyen and Jeng 2021). A recent article in the *New York Times* entitled "A Wall Street Dressing Down: Always. Be. Casual" describes how after returning to the office following the shutdowns of the pandemic the "right" way for men on Wall Street to dress is more casually. There has been a recent shift from wearing suits to wearing casual clothes.

It is not just any casual clothes that will do. The article notes that before going back to the office "some checked with friends to see if their choices were in line with the crowd" (Nguyen and Jeng 2021). This casual wear is as much of a regimented uniform as had been seen previously. And given that it is more revealing of the bodies of the men may also make them feel more social pressure to work out, and have a body that looks powerful without the generous padding of a suit jacket. The uniformity of this new casual dress on Wall Street is evident from the photos that accompany the article and from the discussion of which stores are acceptable to buy these "casual" clothes from. Popular items include

Lululemon's ABC pant, and Untuckit shirts (Nguyen and Jeng 2021). And with time there may be a shift back to the more formal dress.

Because none of this is stated explicitly only the savvy will be wise enough to keep up with these changing social norms. Any young worker hoping to make their way up the corporate ladder would shudder to be the one who is gently ribbed with the "familiar workplace joke: What's with the tie? Got a job interview" (Nguyen and Jeng 2021)? These are the ways the social norms of dress are enforced. With this comment the wearer is given feedback about what he is wearing, told that it is not appropriate and you can guarantee that he will think twice about putting on a tie the next morning. If he knows what's good for his career will promptly make a visit to Lululemon and Untuckit.

Of course, none of this is new. Such attention to the importance of adornment in achieving a place in the upper echelons of society was seen in writings from the time of Darwin as well. In Honoré de Balzac's *Lost Illusions* we see our main character Lucien de Rubempré move from a country nobody to a somebody in the social world of Paris. As Lucien trades his ready-made garments for a Parisian custom suit—at the expense of nearly a year's income—he is able to move up in the social world. And luckily for Lucien he has an "Apollonian" form. As Balzac notes it is not just about the clothes, for in these clothes "men still showed off their bodies, to the great despair of the thin or badly-built" (Balzac 2001). Balzac has an uncanny ability to make social observations that ring true today and here is no exception.

Imitation of Natural Meaning in Non-Human Animals

With imitation of natural meaning, our more base, animalistic selves as bodies—as organisms that are themselves indicators of things to those around us—take center stage. This is no coincidence, and Darwin himself was fascinated by the parallels between bodily "expression" in man and animals. Animals assess each other using various indicators, some that we share and some that are unique to each species.

And animals also lie, or at least, to put it in more neutral terms, convey false information. The conspicuous gold, black, and white patterning on the poisonous butterfly *Danaus chrysippus* is copied by at least thirty-eight other species that are not poisonous (Gregory and Gombrich 1973). An individual butterfly does not have control over its patterning but the outcome is that a potential predator or a rival takes these bodies to mean certain things—things that are not quite true. Animals have bodily features that play off the meaning that is associated with certain patterning, and size. This evolves because it confers an advantage to the species.

Some of the clearest cases of imitation of natural meaning in the natural world are camouflage and mimicry. For instance, in a case of mimicry, the patterning on a stick insect causes the insect to appear branch-like. Through a process of natural selection there were advantages to looking more and more like a stick that made the species evolve in this way (Dawkins 1996). Some non-poisonous snakes have evolved to bear the same patterning as poisonous snakes, which means they are afforded the protection of predators thinking they are poisonous without having to bear the cost of producing venom (Wickler 1968). A certain species of fireflies have evolved to flash the mating signal of another species of fireflies they wish to eat (Skyrms 2010). I will discuss the evolution of non-human animal imitation of natural meaning and mimicry further in the next chapter.

The Evolution of Imitative Adornment

As we have seen, positive imitation of natural meaning can help explain the staying power of the suit, a "uniform so fundamentally successful that its basic elements have endured unaltered for centuries" (Trebay 2014; Stafford 2005: 139). Although adornments cannot evolve through biological sex selection processes because they are not *genetic*, there is reason to think that certain cultural behaviors are selected for through what might still be considered to be Darwinian processes (Dawkins 1976; Dennett 1995; Blackmore 2000; Godfrey-Smith 2011).

Philosopher of biology Peter Godfrey-Smith sees the most promising approach to a discussion of cultural evolution as one where instances of cultural variants, such as "behaviors, psychological states, or material objects" make up "their own Darwinian population" (Godfrey-Smith 2011: 150). In order to form a clear picture of reproduction within a Darwinian population we must "know who came from whom, and roughly where one begins and another ends" (Godfrey-Smith 2011: 86). If we lose this pattern, "in such a way that a vague and disparate set of models all make blended and customized contributions . . . the Darwinian pattern is lost" (Godfrey-Smith 2011: 155). The advocate of a theory of cultural evolution that fits within this parent-offspring framework must answer many complex questions about how parent-offspring relations can be established for cultural entities: questions about the size and form of the replicated units, and what the necessary causal links between "parent" and "offspring" are. The case of bodily adornment—which overlaps nicely with natural, sexually-selected-for features in humans and non-human animals alike—has the advantage (over, say, music) of existing in fairly discrete units with uniform structure. This means that items of adornment seem to be a good candidate for cultural evolution through Darwinian processes. In the next chapter I will elaborate on these connections with Darwin in more detail, and also will further discuss non-human animal bodies.

Imitation of natural meaning has great power to explain a number of behaviors that we engage in, including the persistence of certain garments, such as suits, in terms of how these garments shape what others take to be the natural form of the wearer and how changes in perception of bodies can lead to changes in dress. Communication through bodily adornment also rather neatly aligns with other biologically selected forms of imitation of natural meaning we see in the natural world. Imitation of natural meaning has proven to be a fruitful category for understanding a number of aspects of meaning in bodies and bodily adornment. To conclude the chapter, I summarize the view I've advocated thus far in Table 1.

Table 1 Summary of Types of Meaning

	Natural Meaning	**Imitation of Natural Meaning**	**Non-natural Meaning**
Example	"The fact that Professor Johnson is covered in chalk means she just taught a class."	"His broad shoulders mean he will be a strong leader."	"By wearing the black armband Manuel means 'I protest that my hamster has been banned from the hotel by Basil Fawlty.'"
Detailed by Grice in his 1957 paper "Meaning"	X		X
Meaning of some *thing*	X	X	
Meaning of some *one*			X

Intention recognition needed for uptake			X
Factive	X		
Can lead to implicated content such as irony			X
Can be manipulated or used deceptively		X	X

5

Darwin on Animal Bodies

Introduction

Humans are not the only creatures who have developed bodies whereby certain colors mean things. Just as the red of a shirt on a target worker has meaning so too do the red feathers of a house finch. Of course the matter of intentionality does not apply in the same way to non-human animals. A male house finch is not red because of corporate policy. For consideration of meaning in animal bodies I will now turn to the theories of natural and sexual selection. And although there surely are important differences between what humans and other species can communicate with bodies we also were shaped by those forces—after all we too are animal.

Natural selection is the force behind those physical traits that are passed on from generation to generation. The long necks of giraffes can be explained in terms of the special advantage this provides for reaching the leaves of trees. Some such animal traits can be explained in terms of how they make some creature especially well suited to the natural environment.

However, not all animal traits can be explained in terms of adaptations. Charles Darwin was famously irked by the peacock's tail. It did not help the animal navigate its natural environment, and presents a hindrance to behaviors such as procuring food, avoiding predators, flight, and so on. Some features—such as long, bright feathers—are cumbersome and seem to defy the forces of natural selection. Such physical traits are selected for not because of the particulars of an environmental niche but because of the effect they have on potential mates. These traits arise because of the preferences of members of the opposite sex of the same species, or pollinators of a different species depending on the form of reproduction. To explain features such as the peacock's tail, Darwin developed his theory of sexual selection, published in 1871. Natural selection and sexual selection often push in opposite directions (Dawkins 1996; Andersson 1982).

Understanding this dichotomy between the forces of natural and sexual selection is essential to understanding the ways we use our bodies to convey information. In this chapter I present the intellectual journey behind the development of these theories and provide a number of concrete examples to illustrate the theories along the way. In the subsequent chapter I will turn to explicitly consider the implications of natural selection and sexual selection for humans and will connect the dots with the discussion of meaning in bodies and bodily adornment presented this far in the book.

Natural Selection

Charles Darwin and Alfred Russel Wallace's joint theories of natural selection were presented at the Linnean Society in July 1858 (Raby 2001: 138). One year and some months after this, Darwin's most famous work *On the Origin of Species by Means of Natural Selection, or the Preservation of Favoured Races in the Struggle for Life* was published in November 1859 (Raby 2001: 142). Darwin had been developing the ideas published in *The Origin* since 1837, when it "occurred" to him that "something might be made" of reflecting on observations he had made while sailing the world aboard the *H.M.S. Beagle* (Darwin 1859/2003: 4). Darwin's hand was forced into publishing at this time because Alfred Russel Wallace, a contemporary of Darwin and a natural historian working on the islands of Southeast Asia, sent Darwin his manuscript of a paper containing similar findings and asked for his advice on it (Raby 2001: 137; Prum 2017: 51). Struck by their similarity and wishing to avoid at all cost being thought to have stolen any ideas, Darwin at the suggestion of Joseph Dalton Hooker agreed to have Wallace's paper read along with his sketch at the 1858 Linnean Society meeting (Raby 2001: 141). Wallace was not consulted in this decision because as he was working in the Malay Archipelago, getting his assent would have taken six months and the matter was deemed to have been one of some urgency (Raby 2001: 137–142). Wallace didn't seem to mind, and wrote in his autobiography that "Of course I not only approved but felt that they had given me more honour and credit than I deserved" (Raby 2001: 141). Darwin saw *The Origin* as a "sketch" that he would elaborate on in later publications. After over twenty years of rumination, at the age of fifty, it was time for Darwin to publish his work on natural selection (Darwin 1859/2003; Ruse 2008: 1).

To a philosopher, *The Origin* feels familiar—Darwin presents arguments, comes to conclusions, provides support for the truth of the premises, and considers and responds to objections (e.g., Darwin 1859/2003: 478). Darwin's ultimate conclusion is that the diversity we can find in the natural world can be explained in terms of evolution by natural selection from one origin. The title—*On the Origin of Species by Means of Natural Selection*—can be read as stressing the word "the"—*the* single origin of *all* species, of *all* life, of *all* diversity. We—all species—do not have multitudinous origins but *an origin,* in common.

Darwin's arguments remain controversial in certain quarters even today, 150 years later (Khazan 2019). This is because his theory is often taken to eliminate the need for God as creator. Even today in many parts of America evolution is not taught because, according to political scientist Eric Plutzer, for some educators "their primary concern is not offending the students by characterizing the science in a way that seems to be challenging religious faith" (Khazan 2019). Well aware of these implications, Darwin begins *The Origin* not with the natural world or abstract theories but with a part of nineteenth-century England that would have been familiar to his readers: selective breeding in dogs and pigeons.

Darwin begins the book defensively—by preempting a possible objection to his argument in support of natural selection. How is it possible that all the varieties of life that we find in the natural world—from gorillas to hummingbirds—could have begun from one origin? He provides supports for the premise that species can change. There is evidence, he writes, that all the varieties of dogs we find—terriers, greyhounds, and everything in between—began with one breed that man has bit by bit pushed in different directions (Darwin 1859/2003: 18). He forces a reader to consider: if dogs can change in this way in human hands, then we should not doubt the idea that species in the natural world can change (Darwin 1859/2003: 18). If dogs can change—as the diversity resulting from breeding over a few thousand years shows they can—why couldn't all species?

At the start of *The Origin* Darwin considers pigeons too, a subject he was knowledgeable about himself, as a pigeon breeder (Darwin 1859/2003: 21). This was a pastime common for men of his time (Richards 2017: 177–183). As Darwin notes, based on physical appearance alone, various types of pigeons would appear to be different species: "Altogether at least a score of pigeons might be chosen, which, if shown to an ornithologist, and he were told that they were wild birds, would certainly be ranked by him as well-defined species" (Darwin 1859/2003: 21). All varieties of pigeons were, however, descended from the rock-pigeon. With these discussions of dogs and pigeons Darwin shows our fallibility in classifying species from appearances alone. The take-away from these sections is that some creatures (such as a Yorkshire terrier) can have a vastly different appearance from another creature (such as a bulldog) but be a member of the same species.

If we allow that species can change, do vary enormously, and can be shaped by the selection of breeders, this opens the door to the principle of natural selection. If such changes can happen under domestication, then it can happen in the natural world. To support this implication he writes,

> There is no reason why the principles which have acted so efficiently under domestication should not have acted under nature. In the survival of favored individuals and races, during the constantly-recurrent Struggle for Existence, we see a powerful and ever-acting form of Selection. (Darwin 1859/2003: 486)

On the theory of natural selection the conditions of the natural world mean that those creatures that are the best suited for their environment will be more likely to produce successful offspring. After a number of successive generations this will shape the species in new directions. As co-creator of the theory Alfred Russel Wallace put it: "this self-acting process would necessarily *improve the race*, because in every generation the inferior would inevitably be killed off and the superior would remain—that is, *the fittest would survive*" (Raby 2001: 131, italics in the original). As contemporary research has shown, a feature that leads some creature to succeed in producing 1 percent more offspring would be present in 99.9 percent of the population in 4,000 generations (Pinker and Bloom 1992; Davies 2012: 35).

Natural selection is the process by which animal creatures, plants, fungi, and everything else shaped by evolution gradually develop features that make them more suited to their natural environments. Those features that lead to the most reproductive fitness are those that are passed along in the lineage. As philosopher of science Michael Ruse describes in his introduction to *The Origin*, "Shaggier coats keep sheep warm; stronger legs let the wolf catch the deer; sharper eyes mean that the eagle can spot the rabbit" (Ruse 2008: 3). Stunning adaptations of this sort abound.

Natural Selection and Deception

Natural selection also shapes the messages communicated with animal bodies. In Chapter 4 I discussed the ways that we *Homo sapiens,* such as con artist

Jeremy Wilson, use our bodies to deceive. Similarly, non-human animals use their bodies to deceive and escape being caught. The simplest form of this is camouflage, sometimes called crypsis, where the body of an animal has a similar visual appearance to its environment (Hinton 1973: 100). The back of a yellow-bellied toad is virtually impossible to spot in the mud (Hinton 1973: 99, 118), as is a green lizard on a bright-green leaf. It is not an accident that these creatures have these colors and textures on their skin: these bodies have historically conferred a survival advantage. To avoid being detected by predators is a way of achieving a certain outcome because of bodily patterning, color, or shape. Of course, an individual bullfrog has no control over the fact that he happens to fade into the mud. Perhaps to call this an act of "evasion" is too intentional of a word—but this is the result: the creature evades detection by a predator.[1]

Natural Selection and Mimicry

Sometimes an animal avoids being eaten not by camouflage—resembling the natural world—but by mimicry—resembling another creature. This is seen in the number of snakes that are not poisonous but which resemble poisonous snakes. This resemblance confers an advantage because a creature will avoid eating a non-poisonous snake that appears similar to a poisonous one. This is called Batesian mimicry, named after Alfred Russel Wallace's one time travel companion in the Amazon Henry Bates (Hinton 1973: 99; Mallet and Joron 1999; Raby 2001). In other instances, there will evolve resemblance between "two or more distasteful or harmful species [that] resemble each other," which is called Müllerian mimicry, named after German naturalist Fritz Müller (Hinton 1973: 99).

To put it simply, Batesian and Müllerian mimicry can be understood in terms of survival benefit—although it should be noted that there are complications and paradoxical cases that I will gloss over here (Mallet and Joron: 1999). The advantage of Batesian mimicry is that a creature is conferred the benefit of predators avoiding them without the cost of producing poison or a distasteful flavor (Hinton 1973). With Müllerian mimicry it is less immediately clear why evolutionary forces would lead to the similar patterning. Each creature bears the cost of producing poison or a distasteful flavor, so why mimic others? The answer seems to lie in the memories of the predators: that is, in "predator learning" (Twomey et al. 2013). Birds have remarkably long memories of distasteful and poisonous insects they have encountered (Hinton 1973: 103). Some—such as crows and bluejays—show evidence of remembering these encounters for their whole lives (Hinton 1973: 103). Resembling other poisonous or distasteful insects, even if you are also poisonous or distasteful—is a way to work within the confines of the memories of predators. Müllerian mimicry reduces the cost of predator learning—on the training of the "interpreter" to recognize the meaning of the bodily patterning, if you will (Hinton 1973: 103–4; Twomey et al. 2013).

Deceptive Behaviors

In contrast with bodies themselves that deceive, sometimes a creature will use its body to *behave* in a way that appears to us to be deceptive. Here we see more of a possibility of attributing agency to the creature for the deception. An instance of this can be found in the behavior of birds called piping plovers, an endangered species that lives off the East Coast and in a few select areas of the Great Lakes region (Ristau 1990). Piping plovers perform what has been termed the "broken wing display." Plovers lay their eggs not in trees like most birds but in the sand dunes found in their coastal habitats. This makes them vulnerable to predation from land animals, as well as to being inadvertently trampled on. This unorthodox nest positioning has necessitated the piping plover's development of unusual behaviors to protect their nests.

If another creature—including a human—gets too close to the nest the mother will engage in rather striking behavior. She will position herself at some distance from the nest itself, will squawk and hop around with one wing extended and another wing held close (Ristau 1990). It is almost impossible to miss and it appears that the bird is in distress. Ornithologist Carolyn Ristau (1990) has conducted a number of clever experiments on the plover to test what its broken wing display behavior might indicate to us about the plover's mental capacities and their awareness of other creatures. In particular, she was interested in testing whether the plovers could track the gaze of other creatures in their midst—and, by extension, perhaps the plover's awareness of what the perceived predator was thinking or intending (Baron-Cohen 1995). Ristau's striking experiments showed that piping plovers off the coast of Long Island were more likely to engage in a broken wing display when an experimenter's gaze was directed toward the nest.

With the broken wing display the piping plover uses its bodily movements and sounds to change the behavior of some creature perceived as a threat to their nest. The birds adjust their wings, hop around in an eye-catching way, and make sounds that lead to the attention of the predator being directed away from the nest. They are more likely to act in this way when the perceived predator is looking toward the nest than when he or she is looking away. This is striking behavior, and certainly looks to us like bodily deception.

Sexual Selection

Natural selection can explain many features in the natural world, such as those detailed here, but it cannot explain everything. Those features that seemed to pose a problem for natural selection as fully explanatory irked Darwin. He published his thoughts on such phenomena in his 1871 book, *The Descent of Man*. The twelve-year gap between his publication of *The Origin* in 1859 and *The Descent of Man* in 1871 is at least in part a result of the success that the first publication afforded him (Richards 2017). In the meantime Darwin published more narrowly focused works on orchids, domesticated animals and plants, and so on. Much of Darwin's time in the intervening years was spent lecturing and engaging with the public, as well as in various treatment centers for his persistent bouts of ill health (Richards 2017).

His "abstract" *The Origin* is more to the point than *The Descent*, which at times gets bogged down by examples and anecdotes, both his own and those others have recounted to him. Usually these strengthen his arguments—even if they elongate one small point to an entire chapter—and at other times such anecdotes weaken Darwin's argument because they are hard to believe.[2]

Peacock's Feathers

However, much of Darwin's evidence for sexual selection does not come from potentially apocryphal anecdotes but from features of the natural world that we all can see. Darwin famously wrote that the sight of a peacock's feather made him sick. Peacock's feathers are not like mimicry in snakes, or the broken wing display in piping plovers. Feathers do not afford the peacock any advantage in navigating its environment, procuring food, or avoiding predators. The fact that peacocks carry the additional burden of what amounts to an enormous feathered train seems to present a problem for the very idea of natural selection. Some other explanation was called for. So why did these feathers evolve?

Peacock feathers—and other such features of the natural world—led Darwin to develop his theory of sexual selection. Certain features develop because of sexual selection when they offer some advantage not in navigating the physical environment or avoiding predators of other species, but because they are preferred by the opposite sex. That is, according to Darwin, peacock feathers developed because they were attractive to the females of the species: the peahens. It is the preferences of the peahen that led to this feature that looks so beautiful to us.

Unconscious Selection

There was foreshadowing of the theory of sexual selection in Darwin's discussion of breeding right in the first chapter of *The Origin*. As he describes, a fantail pigeon has an abundance of up to seventeen-tail feathers (Darwin 1859/2003). Darwin writes that these tail feathers are the result of "unconscious selection" at work. According to Darwin no breeder set out to make a fantail pigeon, but rather each successive breeder selected a pigeon with a slightly larger tail; such a tail may have simply made the pigeon look "better" in the eyes of the breeder (Darwin 1859/2003: 34–36). This process of selecting the "more attractive" bird again and again led to the fantail pigeon with seventeen-tail feathers. As Darwin characterizes it there was no "design" or end goal in mind here—just a successive process of selecting the creature that looks "best," which eventually led to dramatic features (Darwin 1859/2003: 34–36; see also Richards 2017: 179). In his later work Darwin makes clear that it is these sorts of gradually changing aesthetic preferences that lead to the features of sexual selection, such as the peacock's tail.

Darwin's theory of sexual selection was not received well by his contemporaries. Alfred Russel Wallace, the co-creator of the theory of natural selection, and the man who provided the impetus for Darwin to publish his results in *The Origin* in 1859, did his best to explain traits such as peacock feathers in terms of natural selection—not sexual selection. Indeed, he and Darwin—who had previously been allies championing the theory of natural selection—parted ways over the issue. Wallace called himself "more Darwinian than Darwin" and seems to have considered Darwin to be a sort of turncoat (Richards 2017: 487; see also Prum 2017: 31–5).

In his 1878 book *Tropical Nature, and Other Essays* Wallace was explicit and public with his position, making it clear that he found the theory of sexual selection "erroneous" (Richards 2017: 486). Wallace maintained the position that natural selection would "itself produce all the results" (Richards 2017: 486; Prum 2017: 34). He preferred instead the "vigor theory," where male ornaments are correlated with "vitality and health" and thus a preference for them can be explained in terms of natural selection alone (Richards 2017: 486). If natural selection is sufficient to explain all we see in the evolved natural world then Wallace would be justified in reaching his conclusion that sexual selection was "unnecessary" (Prum 2017: 34).

Another objection Wallace raised to Darwin was that for the theory of sexual selection to be tenable "all the females within the same district must possess and exercise exactly the same taste" (Richards 2017: 485). Darwin alluded to this in the very first chapter of *The Origin*: the selection is like the "unconscious" selection by breeders for attractiveness (Richards 2017: 485).

What of Wallace's objection that this must mean that all the females exercise the same taste? Modern-day ornithologist Richard Prum defends Darwin's position on sexual selection against attacks like Wallace's. Prum writes that he is in the minority, for, "Today, most researchers agree with Wallace that all of sexual selection is simply a form of natural selection" (Prum 2017: 35; see also Dawkins 1996: 303). Along these lines, philosopher Stephen Davies writes that "theorists now tend to depart from Darwin's view of the separateness of these processes and subsume sexual selection under the broader remit of natural selection" (Davies 2012: 37). The idea in these camps is that the female preference for "bright eyes, glossy plumage," and so on is tracking the male's ability to fight off ever-evolving parasites (Dawkins 1996: 304), which would mean that the preference is not arbitrary and thus not a matter of "mere" unconscious, aesthetic preference.

In his recent work Prum argues for returning to the primacy of sexual selection—and in particular to understanding it as Darwin did, in terms of beauty, a point I will return to in a later chapter.

This remains a controversial part of the theory of sexual selection. A 2009 review article titled "Mate Choice and Sexual Selection: What Have We Learned since Darwin?" calls Darwin's description of a female mallard encountering a male mallard "humorous" (Jones and Ratterman 2009). The article states that Darwin's account of sexual selection is "erroneous," and that the aesthetic component of Darwin's theory of sexual selection "may have been his most significant shortcoming" (Jones and Ratterman 2009). The authors elaborate on this criticism, writing,

> The flavor of Darwin's argument for female choice may represent one of the largest shortcomings of his treatment of sexual selection because it gave the impression that animals would need a human-like sense of aesthetics for sexual selection to operate. Indeed, Darwin himself seemed to subscribe to this point of view, as he went to great lengths to argue that arthropod, insect, and vertebrate females possess sufficient intelligence to appreciate beauty. (Jones and Ratterman 2009: 10002)

Although most modern researchers object to this notion of "beauty"[3] they maintain the idea of female "preference," which sometimes tracks fitness and sometimes does not (Jones and Ratterman 2009). As I will consider in Chapter 9 this may amount to what we think of as an aesthetic sense.

Wallace's main objection to Darwin was that if the preference was merely the "whim" of the female, then the males would never evolve on the basis of this (Richards 2017: 485). For evolution to occur there must be the same choices made over a number of successive generations.

Work first presented by Ronald Fisher in 1915, which was revisited decades later, provides the means to address Wallace's objection (Fisher 1918; Lande 1980; Dawkins 1996: 283–315; Prum 2017: 35–41). We can understand the forces of sexual selection operating on both (1) some feature, such as a bright feather and (2) the preference for that feature (Dawkins 1996: 283–315; Prum 2017: 35–41). According to Fisher, this works in two stages. As Prum explains,

> The first phase, which is solidly Wallacean, holds that preferences initially evolve for traits that are honest and accurate indices of health, vigor, and survival ability . . . in his second-phase model, the very existence of mate choice would *unhinge* the display trait from its honest, quality information by creating a new, unpredictable, aesthetically driven evolutionary force: sexual attraction to the trait itself. (Prum 2017: 36, italics in original)

This process of a mate first preferring a trait because it is an indicator of "vigor," and then later because of a preference for the trait itself was developed by Fisher. Such traits result from what is called a "runaway process" because the preferred trait "runs away" from—becomes no longer correlated with—the feature it initially indicated (Prum 2017: 36).

Preferences for traits and those traits themselves are simultaneously passed on in offspring. If, for example, a peahen has a preference for greener feathers and she chooses to mate with a male with greener feathers, both the greener feathers and her preference for them are passed on in the lineage. The males express and pass on the genetic material for slightly greener tales. Although this preference for green feathers is not expressed in the males, the males carry whatever genetic material is leading to this preference, and will pass it on to their female offspring (Dawkins 1996: 288). The females express the preference genes in their mating behavior, and also possess (although do not express the genes) for longer tails, which they pass on to their offspring (Dawkins 1996). In other words, choosing the greener tail is also choosing for this preference. Thus, "Instead of simply agreeing that females have whims, we regard female preference as a genetically influenced variable just like any other" (Dawkins 1996: 288).

Those females that inherit the preference for green tails also possess the genes for greener tails in male offspring, which will then be combined with similar preferences in their mate, who they will choose on the basis of his tail. This process of the dual evolution of both the sexually selected-for trait and the preference for that very trait is known as "co-evolution" (Prum 2013, 2017). Although I have described it in simple terms here, the viability of these theories of co-evolution in sexual selection is not just theoretical but has been demonstrated by mathematical models as well (Fisher 1918; Lande 1980; see Dawkins 1996: 283–315). So, in summary, Wallace was right to insist that Darwin's theory would require generations of females to have the same preferences, but this has been accounted for within Darwin's theory and developed by later scientists.

Sometimes sexual selection leads to traits that are "maladaptive" from the perspective of natural selection (Lande 1980: 292). That is, sometimes the traits of sexual selection can make it harder for the animal to survive in its ecological niche, to avoid predators, and so on. Brightly colored plumage may make a male bird more appealing to a female, but it also makes it more easily spotted by a predator. Evading predators would lead animals to have camouflage, and indeed we see the result of these pressures on a great number of creatures. That certain animals, like birds, are able to survive with their bright colors shows that being visible to predators does not present a huge cost to the animal. Birds have a way of escaping predators, and it is, of course, their ability to fly away. Female birds also have the ability to fly away but their eggs do not, and this means that if the females were brightly colored—as many males are—then their nests would be in danger. Thus, the females of many bird species are duller in tone than the males.

But the fact that brightly colored male birds can fly away from predators does not mean the feathers come without a cost—it means only that the cost of these colors and long feathers is outweighed by the advantage they confer in attracting

Figure 8 Image of Lady Diana Spencer on the steps of Saint Paul's Cathedral, 1981. We see in this image the exceptionally long train of Lady Diana's wedding dress and veil. The train is being arranged at the bottom of the steps by her bridesmaids, who are so far away they are out of the shot. *Source*: Getty Images.

mates. In his book *The Blind Watchmaker* Richard Dawkins vividly illustrates this point with a discussion of the African long-tailed widowbird. The long-tailed widowbird presents a helpful case because rather than considering the "degree of beauty" of a peacock's tail Dawkins discusses something quantifiable: the length of tail feathers. He writes, "the male long-tailed widowbird is a slender black bird with orange shoulder flashes, about the size of an English sparrow except that the main tail feathers, in the breeding season, can be 18 inches long" (Dawkins 1996: 286). To help you visualize this, the relation of the body of the bird to its tail is about the same as Princess Diana to her train on her wedding day, "the longest in the history of royal wedding dresses," which stretched twenty-five feet behind her (Spranklen 2020; Figure 8).

Dawkins has us consider the evolutionary process that would have led to this bird having tail feathers that are so exceptionally long in comparison to his body. Dawkins has us start by considering "an ancestral bird," with a three-inch tail, "about a sixth the length of the modern breeding male's tail" (Dawkins 1996: 286). He writes,

> Tails have an important job to perform in flight, and a tail that is too long or too short will decrease the efficiency of flight. Moreover, a long tail that costs more energy to carry around, and more to make it in the first place. Males with

4-inch tails might well pull the female birds, but the price that males would pay is their ever less-efficient flight, greater energy costs and greater vulnerability to predators. We can express this by saying that there is a *utilitarian optimum* tail length, which is different from the sexually selected optimum: an ideal tail length from the point of view of ordinary useful criteria: a tail length that is ideal from all points of view apart from attracting females. (Dawkins 1996: 291, italics in original)

Because today the average tail length of the long-tailed widow bird is eighteen inches, we know that this equilibrium continued to move further and further toward the female preferences for long tails. We also know that this did not present such a burden to the males that this cost outweighed the benefit—if so it would not have evolved. We can imagine that in a different context—perhaps in an ecological niche with more predators—the balance between these forces would not have been found here.

Sometimes the result of sexual selection is very costly to the males, and what is costly to males can threaten the survival of the species as a whole. As Lande writes, "In theory, it is even possible for sexual selection to drive a population extinct" (Lande 1980: 300). But for those costly "charms and tools of war" that have arisen, evolution has found an equilibrium point.

On the Matter of Female Choice: "Maleness" Held "Hostage"

Another reason that Darwin's theory of sexual selection may have seemed implausible to his contemporaries is that it puts females in a position of power. As Prum writes in characterizing this type of response, "By diluting the theory of natural selection with a mechanism that rested largely on the power of aesthetic subjective experiences—*vicious feminine caprice*—Darwin had gone beyond the pale of what was acceptable" (Prum 2017: 31). As I will discuss in the following chapter Darwin did see females as selecting for aesthetic features in males—such as bright feathers in birds and for beards in men. However, even up through modern times, over 150 years after Darwin, the idea of female choice is sometimes treated with incredulity.

As Stephen Davies writes in his book *The Artful Species* one reason Darwin's theory of sexual selection did not appeal to his contemporaries was "it made the nature of maleness hostage to the capricious whims of females" (Davies 2012: 38–39). As Fisher's runaway effect and the co-evolution of both trait and preference show there is nothing capricious about these whims, but woefully mentioning female "caprice" is not uncommon (as critiqued in Prum 2017 and Richards 2017, especially 331–369).

Although there is disbelief at "maleness" held "hostage" there is an eager discussion of the female body as an object of male choice to be found in some of the surrounding literature. A favorite topic is female breasts and the ideal hip to waist ratio. For instance, a chapter by Davies on the aesthetics of human beauty begins with a survey of "the literature on youthful female beauty" (Davies 2012: 103). Women's bodies are taken to represent "human beauty," leaving men in the position of "chooser" but not as chosen. Davies notes that although this might be seen as "insulting" and "demeaning" to women, such theorizing can nonetheless "provide a more credible and complex account of human sexuality than its critics suppose" (Davies 2012: 103). It is taken as a given that this would be done from the perspective of a heterosexual man, and discussing sexual selection solely in terms of "youthful female beauty" is not defended beyond this.

When reading discussions of sexual selection by theorists such as Helena Cronin, Anne Hollander, and Evelleen Richards, one notices a very different perspective from that found elsewhere in the literature. Richards is critical of Darwin's "male gaze" and spends a chapter on "putting female choice in its (proper) place" (Richards 2017). Anne Hollander discusses the male body as an object of sexual desire in her enlightening fashion history (Hollander 1993, 1994). And there are, too, some male theorists who have provided an exception to discussing mate selection in terms of female "caprice." Darwin himself, even in 1871, considered men as they were shaped by female choice. This is seen most clearly in his extensive discussion of beards (1871/2004).

More recently Richard Prum in his 2017 *The Evolution of Beauty* includes a lengthy discussion of human penis size as a result of female choice. As he notes, in primates there is a great range of average penis lengths to be found (Prum 2017: 243). For instance, in male gorillas, which weigh an average of 300 pounds, an erect penis is on average one and a half inches long (Prum 2017: 244). The average human penis, in contrast, is four times bigger at six inches long and is also wider than the penis of other primates (Prum 2017: 244). Showing his feminist bent that is apparent throughout the book Prum writes,

> Oddly enough, evolutionary psychologists have not enthusiastically embraced the idea that penis size is an honest indicator of male quality. Although nearly every perceivable feature of the female body—waist-to-hip ratio, breast size and symmetry, facial symmetry and "femininity," and so on—has been scrutinized as a potential indicator of female genetic quality and mating value, the eminently measurable human penis has received little such attention. (Prum 2017: 246)

Another interesting feature of human penises in comparison to other mammals is that the human penis does not contain a *baculum* or bone. Men lost it at some point in our ancestral history (Prum 2017: 246–51). As Prum concludes later in the chapter, in his view the human penis evolved to be what it is today because of preferences for sexual pleasure, experienced both through vision and touch (Prum 2017: 251).

Men are selected for not just because of their bodies but in more recent human history also for the ways they are adorned. We must not lose sight of the fact that our notions of men being more or less adorned or in some specific way are relatively new. When explorers such as Wallace reached Brazil they commented on the fact that it was the men who were more adorned (Raby 2001: 65). This is also found in more recent anthropological literature such as in documentation of the people of Papua New Guinea in the 1960s (Strathern and Strathern 1971).

We should not be led astray by our own cultural expectations of who is adorned in what way. An illustrative case of this point can be seen in an early archaeological finding from 1823 in modern-day Wales known as the "Red Lady." It was thought at the time of discovery that these bones from 33,000 BC which had been covered in red ochre and were found with a number of pieces of ivory jewelry belonged to a prostitute, presumably because of the associations held with the color red and the ornate artifacts (Chapman and Wylie 2016: 2–5). Later bone analysis revealed that the "Red Lady" was in fact a male (Chapman and Wylie 2016: 2–5). Using our own cultural assumptions to interpret past people can lead us down the wrong path.

Fashion history even in the West and just within the past few hundred years shows the enormity of changes in gendered norms of dress. At the time of Darwin, it was a relatively recent development that the adornment of English women had become more ornate than the adornment of English men. In his book *Pretty Gentlemen* historian Peter McNeil writes, "in the eighteenth century women surpassed men for the first time in their pursuit and wearing of rich and elaborate clothing" (McNeil 2018: 18). McNeil also draws on the work of art historian Kaja Silverman, who notes that "ornate dress was a 'class' . . . rather than a gender prerogative from the fifteenth to the seventeenth century, a privilege protected in many European countries and principalities by sumptuary laws" (McNeil 2018: 18).

During these centuries "sartorial extravagance" was not a mark of gender, but a "mark of power" (McNeil 2018: 18; Friedman 2018: 89). The assumption that "fashion" is female and that females are those who are most heavily adorned in Darwin's time and in our own is one that we must not lose sight of.

It is just that: an assumption—and far from a human universal across cultures and times in history.

I will say more about this fashion history and present Darwin's discussion of beards in the next chapter. The disproportionate focus on women's bodies in discussions of sexual selection is something I have attempted to counterbalance by discussing men's bodies here. If theories came from an ideal "neutral" place none of this would be necessary. However, we are not perfectly objective agents. And perhaps it should come as no surprise that our theorizing about bodies is shaped by who has the power in the adorning culture we live in.

6

Human Sexual Selection

Introduction

In *The Descent of Man* Darwin wrote about sexual selection in humans as well as in animals. This is in contrast to *The Origin*, where he avoided the subject of humans, save for a passing reference at the end. One of Darwin's objectives with *The Descent* was to show that all races of people were one species, and that the variation found across *Homo sapiens* was a result of sexual selection (Darwin 1871/2004; Desmond and Moore 2009; Richards 2017). This was a progressive aim, and aligns with how he was raised: Darwin was a part of a prominent anti-slavery family (Desmond and Moore 2009; Richards 2017). But Darwin was still a man of his time in many ways, and he quickly disappoints us if we think he was some sort of saint on matters of gender and race. Darwin held particular vitriol for the people he saw in Tjerra Del Fuego on his voyage on the *Beagle* and saw women as childlike (Darwin 1871/2004; Richards 2017). Despite these major blind spots—and at times outright racism and sexism—Darwin does consider women and those of non-European origins to have agency in the aesthetic preferences by which they select mates and thereby shape future generations. In this chapter I begin with Darwin's view on sexual selection in humans and then present my own discussion of the relation of adornment to gender and race.

Ladies and Feathers

Let's continue in the nineteenth century. As Darwin notes throughout his work, in bird species it is the females who made the choices that led to bird song and bright plumage. And at points in *The Descent* Darwin draws a parallel between the feeling that a female bird would have when gazing on a male bird, and how a woman would feel choosing feathers to put in her hat. This sort of analogy is seen in the following passage:

> When we behold a male bird elaborately displaying his graceful plumes or splendid colours before the female, whilst other birds, not thus decorated, make no such display, it is impossible to doubt that she admires the beauty of her male partner. As women everywhere deck themselves with these plumes, the beauty of such ornaments cannot be disputed. (Darwin 1871/2004: 115)

Women wearing feathers on hats was all the rage at this time in fashion history (Darwin 1871/2004; Richards 2017: 244–52; Wallace Johnson 2019). We see in the quoted text that Darwin drew a parallel between the way a female bird feels when she looks at the feathers of a male bird and the way a woman feels when she considers adorning herself with these very same feathers. Darwin writes that in both cases the woman and bird alike are appreciating the beauty of the feathers.

In her important and incredibly thorough 2017 book *Darwin and the Making of Sexual Selection* Evelleen Richards is critical of this aspect of Darwin. She writes, "Darwin's transposition of women and birds carries with it the sense of female agency in dress and decoration, an agency that is aesthetic in impulse and is designed to attract the sexual attention of human males" (Richards 2017: 244). This is what Richards calls Darwin's "fashion metaphor" writing,

> Through the metaphor of fashion, Darwin might attribute an agency and choice to female birds that he conventionally denied to women. The Victorian middle-class male was preeminently a self-made man—sober, powerful, and purposeful in attire and intent, disdaining display. He could not have been designed by a woman, by female caprice and sexual preference. Having routed the Great Designer from human evolution and racial divergence, Darwin was not about to cede the Creator's place to a woman. (Richards 2017: 256)

Darwin does at points make comments about women that are offensive by today's standards, but he also does something that is perhaps surprising for his time: he considers men as the product of the desires of women.[1]

I believe Richards is right that Darwin may have been reticent to "cede the Creator's place to a woman" as she argues in the passage quoted here. But at the same time, we ought not overlook Darwin's lengthy discussion of beards; Richards mentions beards only passingly in an otherwise very thorough, nearly 700-page book. As Richards mentions, the Englishmen in Darwin's circles saw their beards as a sign of their own superiority over other races (Richards 2017: 134; 152; see also Hughes 2018). In Part III of *The Descent of Man* Darwin begins with a lengthy survey of some of the features of men, including beards, writing,

> Man on an average is considerably taller, heavier, and stronger than woman, with squarer shoulders and more plainly-pronounced muscles. Owing to the relation which exists between muscular development and the projection of the brows, the superciliary ridge is generally more marked in man than in woman. His body, and especially his face, is more hairy, and his voice has a different and more powerful tone. (Darwin 1871/2004: 621).

He continues on the following page, "Man is more courageous, pugnacious and energetic than woman, and has a more inventive genius" (Darwin 1871/2004: 622). Darwin certainly does not sound like a feminist icon with comments such as this, but in beginning this section with a consideration of the physical features of men Darwin does slip out of the "male gaze" that it seems Richards attributes to him.

Indeed, Darwin continues to note that in humans, as in the sorts of bird species he had been discussing, women and children are more similar to each other than men are to women or children. He writes, "Male and female children resemble each other closely, like the young of so many other animals in which the adult sexes differ wildly; they likewise resemble the mature female much more closely than the mature male" (Darwin 1871/2004: 622). As Darwin has argued throughout the book, this difference from the young and female is evidence that sexual selection has pushed the male more than the female to have its distinguishing features.

Darwin continues with his discussion of beards, writing that he knows of no species of monkey where the female has a larger beard than the male, and notes that in man, as in monkeys, when the color of the beard differs from the head it is usually lighter and is often reddish (Darwin 1871/2004: 623–624). He recounts that two men have written to him saying that they are exceptions to this (Darwin 1871/2004: 624). The main focus of the first eight pages of his final part of *The Descent of Man* is devoted to the beards and it is evident that Darwin has devoted much thought and energy to the topic.

Darwin returns to beards later in Part III and explicitly claims that they are the result of female choice. But before he takes this step he turns to consider another male feature that was a result of sexual selection: the greater upper body strength of men. Darwin divided sexually selected-for features into "charms" and those that result from "battle." Physical strength, of course, is helpful in "battle." Reminding us again that he is a man of his time and culture—not some neutral scientific oracle—Darwin begins his discussion by mentioning Helen of Troy and quoting a Latin passage that states, "a woman was the most hideous cause of war" (Darwin 1871/2004: 627). He then writes,

> There can be little doubt that the greater size and strength of man, in comparison to woman, together with his broader shoulders, more developed muscles, rugged outline of his body, his greater courage and pugnacity, are all due in chief part to inheritance from his half-human ancestors. These characters would, however, have been preserved or even augmented during the long ages of man's savagery, by the success of the strongest and boldest men, both in the general struggle for life and in their contests for wives; a success which would have ensured their leaving a more numerous progeny than their less favoured brethren. (Darwin 1871/2004: 628)

As Darwin notes here, he does not take the absurd position that the "broader shoulders, more developed muscles" of men are solely the result of sexual selection, but claims that in addition to these helping in the "general struggle for life" they would also help some succeed in their "contests for wives." Thus, the broad shoulders and muscles of men are at least in part a result of what Darwin calls the "law of battle."

Some twenty pages later in the text Darwin returns to his discussion of beards. He notes that among those races with beards, there is a preference for them (Darwin 1871/2004: 648). He sees this as evidence that "the different races of man differ in their taste for the beautiful" (Darwin 1871/2004: 648). In the case of beards, it would, of course, be the tastes of the women of these races that led to the differences in beards found around the world.

In this section of the book Darwin considers not only the inherited, phylogenetic features of men but also how those are emphasized by grooming behaviors. Speaking again of beards he writes, "It is remarkable that throughout the world the races which are almost completely destitute of a beard, dislike hairs on the face and body, and take pains to eradicate them" (Darwin 1871/2004: 648). As Darwin notes here, the grooming practices of men in those populations that do not have a preference for beards emphasizes these features by shaving. At the time of Darwin, he observed that where beards were thicker (a result of female choice) they were grown and groomed and where beards were thinner (a result of female choice) they were shaved.

The Handicap Principle

As noted in the previous chapter Wallace was vehemently opposed to Darwin's theory of sexual selection. Wallace preferred to understand male features as signs of masculine "vigor," and not as the result of the "mere" preferences of females. A recent incarnation of Wallace's theory, in the work of Amotz and Avishag Zahavi's Handicap Principle, explains beards along these lines.

The Handicap Principle is presented as "a missing piece of Darwin's puzzle" (Zahavi and Zahavi 1997). However, as I hope to have shown, Darwin already had a fully fleshed-out theory explaining features like the peacock's feathers and beards, and it was in terms of female choice and sexual selection. To propose another explanation is not to fill in a "missing" piece of Darwin's theory but to replace an existing one.

The Handicap Principle is the idea that certain types of adornment have evolved because they signal to others that the creature can survive despite the hindrance or "handicap" provided by the feature. On this theory, a peacock's feathers signal to the female his ability to survive even though his mobility is restricted by his tail. On this point Zahavi and Zahavi write, "in order to be effective, signals have to be reliable; in order to be reliable, signals have to be costly" (Zahavi and Zahavi 1997: xiv). In other words, because the tail is costly, in the sense that it makes procuring food and avoiding predators more difficult, it reveals something about the peacock that cannot be faked. For Zahavi and Zahavi the peacock's tail isn't the result of a "mere" unconscious aesthetic preference of the sort Darwin argued for, but a signal that communicates information.

Zahavi and Zahavi explain beards, too, in terms of signaling rather than aesthetic preference, and for them the receiver of this signal is other men. Zahavi and Zahavi discuss beards in a section on the more general topic of threatening a rival by approaching (Zahavi and Zahavi 1997: 17). To threaten a rival by approaching, they write, is to handicap oneself because the male approaching the other puts himself in more danger by the close proximity. This is the cost of the signal and is thus a reliable indicator of the aggression of the one threatening the other. Moving to beards in particular they write,

> In the days before razors, a man's thrown-out chin presented another risk: it brought the threatener's beard nearer to his rival and made it easier for the latter to grab it. By putting his chin out, a threatener shows his confidence that his rival will not dare or will not be able to grab him by the beard or punch him on

the chin—and that he, the threatener, is still confident of winning the fight if the other does dare. (Zahavi and Zahavi 1997: 17–18)

As they see it, to approach another man with your beard was to approach him with a handicap at the bottom of your face—something for him to hold on to in a fight, and doing so would thus signal that you are genuine in your threat. I was surprised to see this explanation for beards because, although I am certainly not an expert in fights between men, I do not believe that in my experience of seeing them in movies, on television, and occasionally in person I have ever seen a man grab another man by the beard as a fighting technique.

Curious about this, I quickly googled and found that there was apparently a well-publicized 2018 National Hockey League incident in which a hockey player named Nazem Kadri pulled off a part of a player named Joe Thornton's beard (Associated Press 2018). The entire first page of my Google search "grabbing beard in fight" was dedicated to this. Kadri reported after the fight that he thought he had grabbed Thornton's jersey, and video footage shows Kadri reaching back for Thornton while bent over by him, so he certainly could not see what he was grabbing (Associated Press 2018). An AP article describes the incident as such,

> "I mean he's a big boy." Kadri said about the 6-foot-4 Shark. "I couldn't reach all the way across his shoulder. I felt like I just grabbed him in the middle of his jersey and just came down with a handful of his hair. I thought I was a hockey player not a barber."
>
> The damage appeared to be done as Thornton lost his balance and went down, with Kadri still attached to his beard. The dislodged hair landed on the ice and the mini-tumbleweed was eventually handed over to the Shark's bench, presumably for safe keeping. (Associated Press 2018)

This case got media attention because of how unusual it was. Coach Peter DeBoer is quoted as saying, "I've seen a lot of things over 25 years of coaching. I haven't seen a clump of beard on the ice before" (Associated Press 2018). As I moved to the second page of my Google search, in an admittedly unscientific analysis (unsurprisingly I did not find scholarly articles on the topic), I came across a website about boxing rules where it was suggested that beards may help in a fight because they (1) cushion the jaw from a blow to some extent, (2) may make it difficult to see exactly where someone's jawline is and land a punch correctly, and (3) cause a punch to slip off your face more easily, especially if the beard is well groomed (Davé 2018). Who knows how accurate any of these explanations are, but given the rarity of beard-grabbing in fights, and the plausibility of beards being an aid rather than a hindrance in a fight, we certainly should not conclude that beards really are an obvious case of a signaled handicap.

Although it is rare to see a beard grabbed in a fight, it is not rare to see long hair on the head grabbed in a fight, especially in a fight between two women. This is something I have seen in movies, on television, and in person. If we consider the handicap theory to be a good explanation then perhaps we should explain the preference men display for long hair in women to be a result of the women signaling that if they got into a fight they could win, despite having this long hair. This is of course highly speculative, but not any more so than the Zahavi and Zahavi discussion of beards in men as an application of the Handicap Principle.

The Handicap Principle can feel convoluted in the face of the simplicity of Darwin's aesthetic preference theory, but, of course, that does not mean it is false. It is certainly counterintuitive that one would signal strength with creating a weakness. Furthermore, in its explanation of beards the handicap theory misses one of the main things Darwin was attempting to do with his theory of sexual selection. Yes, he was trying to explain how features that are glaringly opposed to the forces of natural selection would arise, but he also was attempting to explain the *diversity* that one finds among different members of the same species. This takes us back to his first discussions of dogs and pigeons in *The Origin*. Darwin wanted to explain the diversity of features of one species in particular: *Homo sapiens*. If the Handicap Principle is universal and if there really were a valuable signal sent with a beard then why would some races have beards and others not?

Darwin on Diversity in Sexual Preference

Darwin's theory of sexual selection can help explain the diversity of features found in humans. Tracing the thread of this issue takes us back to Darwin's early years in Scotland, before his expedition on the *Beagle*. The slave trade was outlawed in England in 1807, two years before Darwin was born in 1809, and slavery was fully abolished in England and the British colonies in the 1830s (Desmond and Moore 2009: 12–14). As a sixteen-year-old boy at the University of Edinburgh Darwin trained for an hour a day for two months in the practice of stuffing birds with a former slave from Guyana named John Edmonstone (Desmond and Moore 2009: 18; Riddell 2019). Darwin wrote that Edmonstone stuffed birds "excellently" and recalled that "he gave me lessons for payment, and I used often to sit with him, for he was a very pleasant and intelligent man" (Desmond and Moore 2009: 18). Darwin's family was also an active part of the anti-slavery abolitionist movement, especially on his mother's side, the Wedgwoods—whose name is still known today as the makers of fine china (Desmond and Moore 2009; Richards 2017: 7).

Darwin's most virulent racism was expressed not toward slaves of African descent but for the people of Tierra del Fuego, whom he encountered while visiting the very southern tip of South America on his journey aboard the *Beagle*. On this journey, the British sailors captured Fuegians, brought them aboard, and eventually some were taken to England (Richards 2017). Of these captured Fuegians, one in particular was able to quickly learn English and adopted an English manner of dress and comportment. This Fuegian man was named Jemmy by his captors, who attempted to convert his people to Catholicism and saw themselves as saviors (Richards 2017: 9–30). Darwin and his contemporaries assumed that surely Jemmy would want to stay in England and marry one of their beautiful English girls. They were aghast that Jemmy chose instead to go back to his country and to discover that he preferred his own women to theirs. When he was returned to his place of origin Darwin wrote in correspondence that Jemmy and his wife "paddled away in their canoe loaded with presents & very happy" (Richards 2017: 10), and was utterly baffled by this (Richards 2017: 28–29).

The experience with Jemmy caused Darwin to reconsider his ideas about the role aesthetic preference plays in mate selection (Richards 2017: 29), and he returned to the subject years later in *The Descent*. In the final chapters of *The Descent* Darwin turns to consider beauty and sexual selection in *Homo sapiens*. These musings are based on his own experiences and reports he received from correspondents overseas. And again, as we saw with the discussion of beards from a female perspective, we find in this discussion that Darwin is able to step outside of his own perspective and consider the aesthetic preferences of other people. Darwin writes that he received a letter from a Mr. Winwood Reade in which it was reported that the inhabitants of the West Coast of Africa "do not like the colour of our skin; they look on blue eyes with aversion, and they think our noses too long and our lips too thin" (Darwin 1871/2004: 649). It also was reported to Darwin that the white skin and prominent nose of a British traveler were considered to be "unsightly and unnatural conformations" (Darwin 1871/2004: 646). To their surprise, the faces and skin tone of these British men were not looked upon favorably by those they encountered on their travels.

And, as was reported to Darwin, these British men also learned that their aesthetic preferences in women were not shared by the men they encountered in other places. As Darwin recounts, Mr. Reade told him he does "not think it probable that negroes would ever prefer the most beautiful European woman, on the mere grounds of physical admiration, to a good-looking negress" (Darwin 1871/2004: 649). Darwin also reports that Francis Galton, his cousin and a founder of the disgraced theory of eugenics, remarked that the inhabitants of South Africa were not interested in what he saw as "two slim, slight, and pretty girls" (Darwin 1871/2004: 649), preferring others instead.

Synthesizing these reports, Darwin begins his twentieth and penultimate chapter of *The Descent* with the conclusion that the various races have different aesthetic preferences, writing,

> We have seen in the last chapter that with all barbarous races ornaments, dress, and external appearance are highly valued; and that the men judge the beauty of their women by different standards.... If any change has thus been effected, it is almost certainly that the races would be differently modified, as each has its own standard of beauty. (Darwin 1871/2004: 653)

As we see here, Darwin takes his preceding discussion as evidence that men of different races "judge the beauty of their women by different standards" and thus each has their "own standard of beauty" which would lead to different "modifications" across species.

The surprised reaction to the fact that Jemmy would prefer his own wife over a British woman smacks of more than a hint of racism, as does Darwin's use of terms like "savages," "barbarous," and his alternatingly fetishizing and mocking discussion of "exotic" bodies (Darwin 1871/2004: 644–646). In this discussion of the preferences of different races, as in his consideration of sex, Darwin's shortcomings are apparent. But, at the same time, we do see that he is capable of considering the agency of those unlike him.

In this penultimate chapter of *The Descent* especially and in his work more broadly Darwin considers people of non-European origins as selectors with their own aesthetic preferences, who thereby shaped the physical characteristics of their people. He does not gloss over the fact that the English men were viewed unfavorably in the countries they visited and attempted to colonize (Darwin 1871/2004).

Darwin's discussion of the preferences of the different "standard of beauty" of the races was a part of Darwin's broader aim to establish the unity of *Homo sapiens* as a species. It may be hard to imagine now but at the time of Darwin this was a question that was up for debate (Desmond and Moore 2009; Hughes 2018: 98). The Anthropological Society at the time was "controlled by white supremacists ... who took a fervently pro-South and pro-slavery stance" (Raby 2001: 176). These men "held that white and other races had descended from different stocks" (Raby 2001: 176). In the last section of the penultimate chapter of *The Descent* Darwin briefly considers the differences in skin tone of the different races. He writes that the "best kind of evidence that in man the colour of the skin has been modified through sexual selection is extremely scanty; for in most races the sexes do not differ in this respect" (Darwin 1871/2004: 673). That is, sexual selection is important in explaining sexual dimorphism, where preferences shape the bodies of men and women in diverging ways; skin color is not like this, and so, as Darwin highlights, it appears to be less of a candidate for sexual selection. Darwin does note, however, that "the colour of the skin is regarded by the men of all races as a highly important element in their beauty" (Darwin 1871/2004: 673). Darwin sees some evidence that skin color is a result of sexual preference because it comes relatively late in babies and "the order of development during growth, generally indicates the order in which the characters of a species have been developed" (Darwin 1871/2004: 674). As he notes, babies do not differ in their skin tone as much as adults do, and Darwin sees this as "slight evidence" that these are recently acquired characteristics.

In other words, Darwin suggests that the differences in skin color found around the world are the result of "mere" aesthetic preferences. According to Darwin, skin color is a trait that was only recently developed in our species, not some indicator of vast differences. If Darwin is right, then skin color evolved to be lighter or darker just as a peacock's feather evolved to be a certain shade of green: because of aesthetic choice in sexual selection.[2]

Sexual Selection, Adornment, and Natural Meaning

In his discussions of preferences for certain bodies and the ways that people in different parts of the world have used grooming and adornment to emphasize them, Darwin is presenting instances of what I have called natural meaning and imitation of natural meaning. With discussion of shaving an undesired beard, and applying blush to the cheeks Darwin discusses ways that humans have used the tools and resources around us to imitate natural meaning (Darwin 1871/2004; Darwin 1872/2009: 317). As I discussed in the first chapters of this book natural meaning is meaning others attribute to our natural features, not meaning that is intended to be recognized by others. With imitation of natural meaning we hope that we will simply "look better" to them according to some perhaps unconscious process on the part of the people we interact with.[3]

In his discussion of sexual selection Darwin is focused on those features that are passed on through the genetic lineage. This means that these traits and the preferences for these traits are something that change only over long time spans. This is genetic inheritance and is what led us to having the sexually selected for traits that we possess. Just as we may groom the body in the ways Darwin describes, we also wear different types of clothes that have the same effect. This was not lost on Darwin either. The ways that we choose to groom and emphasize or draw attention to or away from certain parts of our bodies are heavily shaped by culture. We can only change our genetically inherited bodies through many, *many* generations. But preferences for different types of bodies change, at a fairly fast rate, one that genetic transmission could never keep up with. With cultural transmission fashions and grooming practices can change more quickly.

Natural Meaning and Adornment

There are countless ways to use adornment to change the way our bodies look. Wearing the color red will bring out the redness in one's cheeks, similar to wearing rouge. Wearing blue will bring out blue eyes. Wearing yellow will cause Caucasian skin to look wan and darker skin to look more luminous. We can choose to shave or grow our beard, or take supplements, or prescription medicine to make our beard grow thicker. We can focus on our upper or lower bodies at the gym. We can do crunches, squats, bicep curls, and so on. We can choose not to exercise at all. We can pluck our unibrow, bleach our eyebrows white, or grow them thick, tint them dark, or even get them temporarily tattooed on with a process called microblading.

The history of fashion is, among other things, a history of the ways preferences for different bodies have changed over time. As I noted Anne Hollander convincingly argues that the staying power of the suit—which has persisted for over 200 years—can be explained in terms of the way it changes the natural appearance of men's bodies. As she argues, because our idea of the ideal male form has not changed much in the past 200 years neither has this garment.

Female Preference for Certain Traits

This discussion of imitation of natural meaning in humans might naturally lead one to wonder if this is also found in non-human animals.[4] Do non-human animals alter the appearance of the bodies they inherit? Now, of course, animals do not wear dresses and suits, but if they could do the equivalent of this, would they? In a 1982 study of the long-tailed widowbird, researcher Malte Andersson conducted an experiment that hints at an answer. The researchers had observed that those male widowbirds with longer tails had more mates. This made them wonder what would happen if they adorned the birds in such a way that this feature was artificially elongated. To answer this question Andersson and colleagues actually glued feathers to some of the male widowbirds, artificially shortening or lengthening them (Andersson 1982; Dawkins 1996: 304–6). Would this have an effect?

Would the females be able to detect the meddling with nature—the "deceptive" imitation—and be put off by this? To account for this possibility Andersson included two control groups: one that had no changes at all, and one that had feathers that had been removed but then adhered in the same way, without any lengthening or shortening, and found no change (Andersson 1982). Andersson then waited to see how many nests containing eggs were in each male's territory. It turned out that those males with the elongated tails attracted almost four times as many females as the males with the shortened tails (Andersson 1982).

We can understand this experiment in terms of imitation of natural meaning. Just as we alter the appearance of our bodies with suits and Just for Men with the aim of achieving a certain effect, we see that the male long-tailed widowbird is treated differently by females on the basis of his tail length. We do not know if the artificially elongated tail of the male long-tailed widowbird is read as a signal for some sort of fitness or if it is a "mere preference." But it is deceptive in the sense that the females who mate with these males with the artificially elongated tails will not produce offspring with longer tails, although their offspring will inherit the preference of the mother for this trait. In Andersson's experiment we see that imitation of natural meaning in male long-tailed widowbirds led to more reproductive success.

Given that this experiment demonstrates that the females have a preference for longer tails, why then would the forces of sexual selection not have pushed the tails to be longer? The answer lies in the interplay between sexual and natural selection. Although the female long-tailed widowbirds have a preference for a tail of a longer length than the males have, natural selection cannot sustain this. The result is that the males have shorter tails than the females would prefer—even at the extreme of eighteen inches. The two forces of natural and sexual selection have led to an equilibrium that balances the cost in the natural environment with how much more this causes a female to favor some particular male (Dawkins 1996).

Adornment and Co-evolution

As Darwin argued, many of the features found in the races around the world are there because of sustained preferences for these traits. This is because the trait and the preference for the trait are both passed on in the genetic lineage. In other words, a trait and a preference for a trait co-evolve.

Co-evolution happens both for our inherited traits and the ways that we choose to adorn ourselves because of more passing fashions. On this topic, while discussing Darwinian sexual selection, Prum draws a connection between the process of co-evolution in animals and in human fashion. He writes the following about co-evolution,

> This evolutionary mechanism is rather like high fashion. The difference between successful and unsuccessful clothes is determined not by variation in function or objective quality (really) but by evanescent ideas about what is subjectively appealing—the style of the season. Fisher's model of mate choice results in the evolution of traits that lack any functional advantages and may even be disadvantageous to the displayer—like stylish shoes that hurt one's feet, or garments so skimpy that they fail to protect the body from the elements. (Prum 2017: 40)

We see the same process of co-evolution in bodies as well as in bodily adornment. As Prum suggests in this passage, as some type of garment, say, crop tops and high waist pants, becomes more popular, so, too, does the preference for this trait. In certain cases, some garments are a mere blip on the radar because a small segment of the population chooses to wear them but a preference for the garment in others does not grow. Of course, the means by which bodies and adornments are replicated happen on completely different timescales; nonetheless the underlying co-evolution of both trait and trait preference is found in both.

Natural and Sexual Selection in Bodily Adornment

As we have seen, the forces of natural selection and sexual selection can push in opposing directions. Natural selection alone would push toward neutral-toned birds that could best avoid predators with efficient feathers that are optimized for flying. In the case of the peacock sexual selection has outpowered natural selection and created instead excessively long and brightly colored tail feathers.

Natural selection would lead to traits that make the creature best suited for the environment; sexual selection leads to traits that are appealing to mates, and which are often an inconvenience in facing the natural world (Milam 2010). These two forces also are at play in our bodily adornment.

Our clothing choices are more complicated than what could ever be simply broken down into choices for suitability in either the natural world or to attract a partner, but these are often found in opposition. For those in very cold climates, for many months out of the year there is a tension between wanting to wear something that "looks nice" and wanting to wear something that will be warm enough. Wearing a bulky warm coat, hat, scarf, gloves, boots, and two pairs of pants does nothing positive by way of imitation of natural meaning.

In Darwin's time, the adornment of those encountered in other parts of the world was taken to be indicative of the character of the people—a sign of their inferiority to the British, and something to be rectified. Dressing Jemmy in his English garb was seen as taking a "savage" and "civilizing" him. As these people were colonized the hope and often explicit aim was that they would grow to be ashamed in their bodies. There was an aim toward "civilizing those naked peoples by coercing them into wearing clothes" (Gilligan 2019: 6)—often cloaked by talk of "progress." This was seen not as a matter of mere decoration but of morality.

But as these British explorers discovered, many of their clothes were not suited for the environments of the Amazon or the Malay Archipelago. Just as cold weather necessitates wearing a coat, these environments shape adornment. As Wallace discovered, there was a reason that the people of Serra de Cobati went naked through the woods—his clothing would constantly get caught on trees (Raby 2001: 62). Historian Peter Raby writes that "Wallace found his clothes and equipment a nuisance: his gun would catch on overhanging branches, and the hooked spines of the climbing plants caught on his shirt-sleeves, or knocked his cap off" (Raby 2001: 62). One can only imagine how foolish this must have appeared. Perhaps akin to someone who wears shorts to ski slopes, or flip flops to ride a horse. Why would this strange white man insist on wearing these things that are so clearly not suited to the environment and which slowed him down? Wallace—who had a degree of self-awareness that allowed him to learn from the native people he encountered—wrote, "The Indians were all naked, or, if they had a shirt or trousers, carried them in a bundle on their heads, and I have no doubt looked upon me as a good illustration of the uselessness and bad consequences of wearing clothes upon a forest journey" (Raby 2001: 62). Wallace recognized some of the foolishness of himself and the other British colonizers in these moments. He also noted the hypocrisy of missionaries who had children out of wedlock and remained skeptical of their "imposition of Christian ritual on the local people" (Raby 2001: 65).

In most cases we wear garments that are suited for the natural environment—although sometimes ideas of "propriety" and modesty get in the way, as for Wallace, who would have had an easier time tracking birds in the nude. And sometimes a garment is fashionable but not suited for the natural environment. Prum notes that "the style of the season" may lead to "garments so skimpy that they fail to protect the body from the elements" (Prum 2017: 40). This is not simply a seasonal issue but a perennial fact for women. Speaking very generally and sweepingly, at formal occasions—such as a state dinner—men will wear a suit or a tux with a tie, with socks, shoes, and nearly their entire form covered, save for their hands and head. At formal occasions women will usually wear a dress with differing parts of their body exposed—nearly always their arms—depending on the dress and the occasion. Socks will never be worn by women in formal attire (except on the occasions where a woman is flouting the gender norms of dressing). If one notices attendees at formal occasions, or perhaps notices photos taken at after parties, the practical begins to take precedence over appearances. Women will take off their high heels. Men will take off their jackets and ties, rolling up their sleeves. A woman might end up wearing the jacket of a man over her dress. The sexual dimorphism of fashion at the start of the night begins to blur slightly.

We see something similar in office environments, where women are also often cold, sometimes hiding sweaters or even blankets and space heaters in their offices or at their desks. When the temperature in offices is set by men in suits and socks it will be far too cold for women in skirts and without jackets. In 2018 Cynthia Nixon and Andrew Cuomo, both then contenders for the Democratic governor's seat, had a disagreement about the temperature of the debate hall at their upcoming event (Pager 2018). Nixon said Cuomo's desire for the hall to be cold was sexist (Pager 2018). In a 2015 study showing the lower resting metabolic rate of women, the authors argued that office temperature settings exclude women (Kingma and van Marken Lichtenbelt 2015; Belluck 2015). They argue that the reason many women in offices are cold and need to have blankets in their offices is a combination both of their resting body temperatures and the gendered norms of dressing.

Just as women are bound to wear clothes that make them cold on formal occasions, and in offices, men are often bound to wear clothes that make them hot on certain occasions and in certain environments. Although this conservatism may be changing slightly with time, a man who works in a white collar job cannot wear shorts or a sleeveless top to work, or to a dinner party, or to his wedding without flouting gender norms. Former President Barack Obama was famously excoriated for the innocent act of wearing a tan suit one hot August day in 2014 (Noori Farzan 2019). Just as a woman may resent being cold in an office (Kingma and van Marken Lichtenbelt 2015; Belluck 2015), men may resent the rigidity of standards of dressing that do not permit them to wear shorts or sleeveless shirts to work in the summer. Such garments are simply not considered to be "professional" by the majority.[5]

The fact that suits have persisted as the standard of "professional" menswear may have something to do with the effect they have on the natural meaning that is attributed to men's bodies. Donald Trump is not known for his figure, but in a suit his body is shaped in a way that becomes clear if one has seen a photo of him in more casual attire. A suit allows a man of advancing age to appear "dignified" in a way other clothing cannot. And given who is in power in the business world is it any wonder, then, that suits are the standard item of dress in many professional environments? And that when women began to enter the workforce in greater numbers in the 1970s and 1980s that this is the look they emulated (Hollander 1994)? In adornment, as with bodies themselves, we see a number of conflicting forces are at play.

School Dress Codes

Who has power in society has long shaped what sorts of garments are deemed acceptable. Historically, we see long traditions of attempts to control and constrain the body of women—to dictate what they can wear and when. Throughout history and across cultures, women have been alternatingly praised and shamed for what they are wearing. Like many women, I have had personal experience with this. One case that stands out in my memory is being called to the assistant principal's office in High School for wearing a tank top. This was the first and only time in High School I was sent to the assistant principal's office—and it was so my body could be policed. Tank tops did not seem inappropriate to me then and do not seem so now. And as I pointed out to in the assistant principal's office, my school did not have a rule against them. I was informed they were planning to make one. Teen girls are presented with constant conflicting messages about their bodies but there is something especially pernicious about the policing being done by their school, with all the gravitas that this holds.

The Stonewall Riots

The way we are expected to use adornment is gendered. Men can go topless at the beach and, in most places in America, women cannot. Women are sometimes policed for showing "too much" of their bodies and other times for being too covered up or for not being feminine enough. In some countries, the gendered rules of dressing are even more specifically mandated by law, and until recent history gendered dress codes were widely mandated in America. In the 1969 Supreme Court case *Tinker v. Des Moines* dress was deemed by the United States Supreme Court to be protected as speech. At the time, in many locales there were laws about the number of "masculine" items men must wear and the number of "feminine" items women must wear (Carter 2010).

It is just these laws that—in addition to other reasons involving the mafia and NYPD corruption—led to the famous Stonewall Riots in the summer of 1969 (Carter 2010). On the night of the Stonewall Riots members of the NYPD went into the Stonewall Bar on Christopher Street in New York City and started to arbitrarily "enforce" the law about what men could wear and what women could wear. Their brutal "enforcement" included forcibly bringing bar patrons into the bathroom to "check" their sex—forcing them to reveal their genitals to the officers. The other bar patrons revolted and this led to rioting, which continued for a number of days (Carter 2010). These riots helped spark the gay rights movement and this history is the reason that LGBTQ pride parades around the world happen in June and July, in commemoration.

Nightclub Dress Codes

Our bodies are policed—literally as in the Stonewall Riots—and sometimes in slightly more subtle ways, such as by schools when we are children, or by bouncers when we are adults. In his work on nightlife dress codes, sociologist Reuben A. Buford May draws attention to the way that nightclub dress codes disproportionally affect African American men (May 2014: 69). He writes that the styles associated with hip-hop culture such as "athletic jerseys, baggy jeans, over-sized plain white T-shirts, sweatbands, do-rags (polyester head wraps), 'wifebeaters' (tank tops) and thick gold chains" are those most likely to be prohibited according to nightclub dress codes (May 2014: 71). Buford May writes that these night club dress codes reflect "the racism and classism of their patrons," because "if key clientele were willing to frequent nightclubs where class and racial distinctions were not important, nightclub owners would have little economic incentive to institute policies, like dress codes, that support exclusion of particular types of patrons" (May 2014: 89). The not-so-subtle message that is conveyed by dress codes such as "no wife beaters, no gang wear, no saggy pants, and no flat bill hats" (May 2015: 40) is that African American men are not welcome—unless these potential patrons will conform to what might be seen as "dressing white."

Reappropriation and Metalinguistic Negotiation

We see in the cases of the Stonewall Riots, and dress codes at schools and nightclubs that certain bodies are policed more than others. And throughout my discussion of natural meaning in bodies I discussed the ways that bodies are taken to naturally *mean* things—not because anyone intends to mean anything but because *bodies themselves* are taken to have this meaning. But where does this come from? Who sets these expectations? Can we change them? Should we?

For discussion of this point it is again helpful to begin with philosophy of language, and move from there to reflexively discuss adornment. To consider the point about change in the meaning of adornment, let us now consider by way of analogy change in word meaning. Word meanings change over time and have a sort of "natural drift" of meanings, if you will, where they change by shifting "like a cloud" (McWhorter 2003). "Going to" will change to "gonna." "You all" changes to "y'all." Most times, this change is slow and somewhat arbitrary, done without specific aims. But sometimes, for certain words, there is an attempt to consciously change what they mean—in some instances, to "reclaim" them, and in doing so empower those who have a stake in their meaning. For instance, the word "queer" was once viewed as a slur but now has been reappropriated by the community. By embracing this term, the queer community was able to denude it of its power.

Reappropriation can happen through a conscious process of what is called by philosophers "metalinguistic negotiation" (Burgess and Plunkett 2013; Plunkett 2015; Thomasson 2017; Cantalamessa 2020). On this view, philosophical disagreements which otherwise may appear substantive, are in fact are expressing instead a position about how *terms and concepts* such as "'person', 'free will', existence', and 'artwork'" should be used (Cantalamessa 2020: 124). As philosopher Elizabeth Cantalamessa writes in her recent work on metalinguistic negotiation and art, "such debates are better understood as *advocating* for certain conceptual schemes rather than *reporting* empirical discoveries" (Cantalamessa 2020: 125) about people, free will, existence, artwork, and so on.

How we understand philosophical terms can change how we view the world, particularly if we are engaged in the practice of philosophy. "How terms are used is important because they are subsequently incorporated into our thinking and reasoning and so have direct consequences on how we think about and interact with both ourselves as the world" (Cantalamessa 2020: 125). It is true that for many people the world will seem different depending on their understanding of persons, free will, existence, artwork—such is the value of philosophy—but not everyone has reflected on these enough to have a stake in the matter.

Perhaps the categories that most affect our lives and our understanding of reality are those that describe the groups of which we are members. This is why a queer person would have a particular stake in how the word "queer" is used. Through using metalinguistic negotiation to reappropriate the term a queer person can change "how we think about and interact with both ourselves and the world."

But the "target group" are not the only ones who have a stake in how some term is used. This is because, as James Baldwin has noted, people often define themselves in contrast to others (Baldwin 1960). In a 1960 interview with CBC Baldwin said the following about racial categories:

> In a way black men were very useful for the American, because, in a country so absolutely undefined—so amorphous, where there were no limits, no height really, and no depth—there was one thing of which one could be certain. One knew where one was, by knowing where the Negro was. You knew that you were not on the bottom because the Negro was there.

Because of the point that Baldwin makes here, non-queer people, may have a stake in how words like "queer" are used, because it can be part of one's identity that they are *not* queer (see also Barnes 2016).

And when we are definitely members of some community—African American, gay, woman, disabled, whatever it may be—we bifurcate these further. Author and advice columnist Dan Savage has described how when he was growing up he saw a particular representation of a gay man on the television show *Barney Miller*, "very swishy, total stereotype, carried a purse, owned a poodle" (Savage 2007). Savage reports that he knew that he was gay, but was determined not be "that kind" of gay man—who he calls simply and derisively "the swish"—although in a twist of irony he later in life ends up being a poodle-owner, something he has to work to come to terms with (Savage 2007).

Crinolines and Slutwalks

Along these same lines, young girls in America are raised with explicit and implicit messages that although they are inextricably women, they cannot be a *certain* kind of woman. It is this sort of mentality that is behind laws and policies about nudity, school dress codes, and also underpins other, more subtle attempts to control what women wear. And, of course, as we have seen again and again with words as well as adornment, the norms surrounding what is thought to be "proper" for a woman to wear have changed over time.[6]

As Alison Lurie points out, shame about the body being improperly adorned is the first reported experience of humankind in the Genesis story of the Garden of Eden. As Adam and Eve realize they are naked, they feel shame, and sexual "modesty" is the reason for the invention of clothes. It was in realizing that they ought to be ashamed that they "sewed fig leaves together, and made themselves aprons" (Lurie 1983: 212). In the Bible story, shame about the body was humankind's first lesson and curse.

We see throughout history different notions about what degree of visibility of the body is shameful and what is acceptable. There are certain things that were worn in the past that would be deemed unacceptable in contemporary society. We know the first Olympics in Ancient Greece were run naked, and during the Amarna period in ancient Egypt women wore dresses that were made of very thin, transparent cloth, of the type that would get you arrested in most places in America (Robins 2008). Corsets, which in our time have a sexual connotation, were in the 1830s thought to be a necessary requirement for any proper lady (Lurie 1983: 217). Indeed, a "proper" lady in the nineteenth century might wear anywhere from ten to thirty pounds of clothing (Lurie 1983: 218).

In contrast, certain pieces of adornment from relatively recent history that to us appear very conservative were considered to be improper at the time. The nineteenth-century long, thick crinoline skirts, that to us look very covered and "old fashioned" were thought to be the shameful clothing of the day (think Scarlett O'Hara's skirt).

Then, as now, we see other women doing the brunt of the policing work of this behavior. In her 1883 publication "The Girl of the Period," which was "dedicated to ALL GOOD GIRLS and TRUE WOMEN" E. Lynn Linton writes that for the scandalous modern girl,

> Her main endeavor is to outvie her neighbours in the extravagance of fashion. No matter if, in the time of crinolines, she sacrifices decency; in the time of

trains, cleanliness; in the time of tied-back skirts, modesty; no matter either, if she makes herself a nuisance and an inconvenience to everyone she meets. . . . Nothing is too extraordinary and nothing too exaggerated for her vitiated taste; and things which in themselves would be useful reforms if let alone become monstrosities worse than those which they have displaced so soon as she begins to manipulate and improve. If a sensible fashion lifts the gown out of the mud, she raises hers midway to her knee. If the absurd structure of wire and buckram, once called a bonnet, is modified to something that shall protect the wearer's face without putting out the eyes of her companion, she cuts hers down to four straws and a rosebud, or a tag of lace and a bunch of glass beads. (Linton 1883: 3)

We see in this diatribe how arbitrary the standards applied to propriety are. We can learn from this passage that crinoline skirts were deemed indecent, trains unclean, tied-back skirts immodest. Who would have guessed that a small hat with "four straws and a rosebud" or with "a tag of lace and a bunch of glass beads" would be objectionable? By looking at historical examples we can see how mutable all of these meanings attached to such garments of "unsensible women" really are.

Today's norms of dressing for women may feel relaxed in comparison to the picture painted by Lynn Linton. Women wearing pants is one stark example of recently changed views about what is acceptable for women to wear. It was in the 1890s, with the invention of the bicycle, that women started wearing a precursor to pants, or as they were called as the time: a "divided skirt" (Lurie 1983: 225). In the 1920s it became acceptable for women to wear pants for sports and lounging. In our modern society a woman wearing pants is far from a sexual or shocking thing. Indeed, in a turn of events that might shock Lynn Linton, pants are thought to be a less sexual option for women. This is the reason that pants are worn by many women politicians. In fact, *not* wearing pants as a modern woman politician is now thought to be bucking the norm (Friedman 2018).

Today it is not shocking to see a woman wearing pants in almost any context—except for maybe one. We have recently seen modern heterosexual women wearing "bridal pantsuits." One Spring 2020 article starts with the line "Take a seat, traditional wedding gown—these days more and more brides are opting for pantsuits" (Harano 2020; Pennell 2020). A bridal pantsuit pushes modern boundaries and might surprise and possibly even shock or offend some wedding guests. Perhaps this shock is analogous to how the first viewers who saw a woman riding a bike in a divided skirt felt.

Within the past 200 years Western women have done what they could to liberate themselves from whatever norms of dressing seemed to be holding them back. This attempt at liberation continues with a modern case I understand to be an instance of metalinguistic negotiation with dress. As described in the previous section, recall that metalinguistic negotiation is a means of taking a normative stand about how terms and concepts *should* be used (Burgess and Plunkett 2013; Plunkett 2015; Thomasson 2017; Cantalamessa 2020). With metalinguistic negotiation in bodily adornment the idea is that through certain acts of flouting norms of dressing one can implicate that the meaning attributed to a certain way of being or dressing ought to be changed.[7] But this is easier said than done.

SlutWalks

Events called "SlutWalks" have taken place across America and around the globe since 2011. The SlutWalk events were first sparked by the remarks of a Canadian police official, who, in responding to a number of rapes at Canada's York University, said that "women should avoid dressing like sluts in order not to be victimized" (Valenti 2011; McAfee 2015). SlutWalks began as a direct response to the idea that a woman is to blame for her rape because of what she was wearing. The attendees at these events would usually wear revealing clothing, lingerie, or to various degrees be nude. The organizers know that the dress of the protesters and the event's name would shock—just as it perhaps shocked when there were the first steps taken to reclaim the word "queer," a connection that was explicit (Morrigan 2015). The attitude taken by the protesters is "you may think my wearing this makes me a slut, but I will reclaim my body and these garments for myself, and thereby show you that you need to reconsider your conceptual framing of myself and my body." As characterized in the *New York Times*, "SlutWalkers want to drain the s-word of its misogynistic venom and correct the idea it conveys: that a woman who takes a variety of sexual partners or who presents herself in an alluring way is somehow morally bankrupt and asking to be hit on, assaulted, or raped" (Traister 2011). The organizers and protesters know that this way of dressing flouts norms of what is acceptable— just as the name of the event does—and it is in these acts of flouting norms that the act of protest occurs.

Journalist and commentator Jessica Valenti writes, "Thousands of women—and men—are demonstrating to fight the idea that what women wear, what they drink or how they behave can make them a target for rape" (2011). The idea of this event took off, and in 2011 Valenti reported that "SlutWalks started with a local march organized by five women in Toronto and have gone viral, with events planned in more than 75 cities in countries from the United States and Canada to Sweden and South Africa" (2011). She declares that "In just a few months, SlutWalks have become the most successful feminist action of the past 20 years" (2011). These events gained a lot of publicity, perhaps some due to the message but also because they involve women dressed in revealing clothing, and thus also drew the intended target for change: those who view the marches with a lascivious gaze.

Although Valenti declared this to be "the most successful feminist action of the past 20 years" in 2011, criticisms of the marches came almost as quickly as the praise, with approximately a quarter of articles published about the movement being commentary that denounced it (Mendes 2015). One criticism is that "the package is confusing and leaves young feminists open to the very kinds of attacks they are battling" (Traister 2011). This same observer and critic writes,

> I found myself again wishing that the young women doing the difficult work of reappropriation were more nuanced in how they made their grabs at authority, that they were better at anticipating and deflecting the resulting pile-on. But I also wondered if, perhaps, this worry makes me the Toronto cop who thought women should protect themselves by not dressing like sluts.

We see that in one article SlutWalk is objected to for being both confusing and also not nuanced enough. Nuanced messages are harder to get across. And as we see here, even the critic acknowledges the possibility that perhaps she is just uncomfortable with the bucking of norms that is necessary for change. However, such criticism misses the point. The confusion that results from the flouting of norms is a part of how the process of changing meanings works: it is what makes it clear that metalinguistic negotiation is happening—that the protester is using the term or garment in a new way.

In 2014 the name of what had formerly been SlutWalk changed in many marches to the "March to End Rape Culture" (Weymouth 2014). One commenter protested to this sanitized version of the name writing, "I can't help but think that it's lost some of its power." Some explained the change in name as resulting from organizers having "caved to feminists who reject the word and officials who think it invites trouble" (Weymouth 2014). Organizers in Philadelphia said they changed the name because a number of participants said that they did not relate to the word "slut" (and that for them the word "ho" was more damaging) (Weymouth 2014).

Although these reasons are worth noting it is unfortunate that the name of the event became one that no longer flouted norms of politeness, and with the name change to something inoffensive a parallel was lost with the flouting of norms of dressing. Although from the March to End Rape Culture website it appears that some women still march at these events in some degree of undress, without this context the attempt at metalinguistic negotiation with the adornment seems to have gotten lost. Despite this, there do still remain a number of ways that expectations with dress are flouted, from women wearing pantsuits at their weddings, to the very first women who wore "divided skirts" on bicycles. Sometimes it is important to push back against the meanings that are attributed to our bodies and our adornment, but, as these cases demonstrate, it is not easy, and guardians of old norms hold to their convictions with resolve.[8]

7

The Evolution of Bodily Adornment: Signaling and Meaning-Making in Prehistory

Introduction

How did adornment ever get this power in our lives? In this chapter I turn to consider bodily adornment as an evolved behavioral practice. I have been arguing for taking bodily adornment seriously as a conveyer of meaning. It is not something trivial or simple. This is never more apparent than when we consider how important bodily adornment is for the archaeological record, and for our understanding of how we as a species developed systems of symbolic meaning.

The ability to convey intentional, symbolic meaning has sometimes been presented as what distinguishes us from beast. Archaeologist Erella Hovers has written that "For many researchers the ability to create arbitrary relationships between ideas and their referents—that is, to construct and use complex symbol systems—is the defining characteristic of *Homo sapiens*" (Hovers et al. 2003: 491). Certainly this has been breathlessly argued about remains such as early cave art. But, in fact, the earliest proposed symbolic practices in the archaeological record are not cave art but bodily adornment.

As with the previous cases I have discussed, in this chapter I will argue that adornment in prehistory is best understood in terms of natural and non-natural meaning. I will discuss our earliest symbolic behavior, found in the form of ochre that adorned the body and in shell beads. As we will see, meaning in bodily adornment plays a key role in understanding who we are, where we came from, and our uniquely human practice of engaging in certain forms of symbolic behavior.

Imitation of Natural Meaning in the Middle Ages

As we have seen with more recent cases such as the suit, imitation of natural meaning changes as our ideas about natural meaning and the body change. According to Anne Hollander, this is the reason that the suit became and has remained prominent. As new types of bodies gain positive associations, garments that imitate those natural features will emerge.

At certain points in history we see trends that have come and gone. Some of these stand out to us as bizarre because they rely on an association we no longer have, or imitate a feature that we now have negative associations with. For instance, in medieval paintings we can see that women used to pluck the hairs at the front of their hairline creating a strikingly large forehead. This was thought to be beautiful, a sign of refinement (Figure 9). Additionally, in paintings of nude women underarm and pubic hair is also removed (Friedman 2018: 81). The prevalence of this practice is further supported by textual documents

Figure 9 *Portrait of a lady*, by Rogier van der Wayden, c. 1460. In this painting we see a woman who is beautiful by today's standards except for her hairline, which would have been plucked to look like this. It is an instance of imitation of natural meaning that had positive connotations at the time. *Source*: National Gallery.

such as Latin cosmetics manuals from the twelfth to sixteenth century, which included instructions on how to remove body hair (Friedman 2018: 81-2). In medieval times female body hair was seen as a sign of "probable infertility" and "domineering behavior" (Friedman 2018: 95), and thus for reasons ranging from "vanity to fear of marital rejection" body hair was removed (Friedman 2018: 81).

This grooming practice stands out to us today because the hair was plucked so that the forehead extends nearly to the top of the skull. This comes across as unsightly mutilation of women who are otherwise beautiful by modern standards. Given this deviation from our own beauty standards, these are the sorts of cases of imitation of natural meaning that are easiest to identify from the vantage point of today.

Imitation of Natural Meaning in Ancient Egypt

Moving back further in history, in sites and artifacts from ancient Egypt we find a wealth of depictions of nobility. In ancient Egypt we can again find clear evidence of imitation of natural meaning in adornment. In statues and bas relief we see kings and other leaders are depicted in such a way that they convey power to us today. A wide stance, with narrow hips, little body fat, and broad shoulders, to us looks powerful and is close to the same "male ideal" portrayed by the ancient Greeks. This is the same male ideal that, according to Hollander, we still imitate today with the suit. For thousands of years this was a relatively consistent style of depiction in Egyptian art (Robins 2008). We can see the power in these statues and other depictions despite the gap of thousands of years between the intended audience and us as interpreters.

It is against this backdrop that the art of the Amarna period stands out. During this period, from 1353 to 1336 BC, there was a change in religion, a change in the place of residence of the king, and a striking change in the ways bodies were depicted in the art (Robins 1993, 2008). As with the modern suit, we find a change in the ideal body leads to the emergence of garments that emphasize different features of the body. This is evidence of imitation of natural meaning.

The Amarna period is marked by the reign of King Amenhotep and the famous Queen Nefertiti, and overlapped to some extent with the short reign of their son, the young King Tutankhamen (Carter 1922/1977; Robins 1993, 2008). In this period, there was a shift from worshiping the solar deity Ra-Horakhty, who was depicted in human form with the head of a falcon, to worshiping the sun god Aten (Robins 2008: 149). In the art of the time Aten is represented as a

Figure 10 Statue of King Akhenaten at Karnak, c.1365 BC. In this statue we see King Akhenaten depicted in the Amarna style with no visible genitals and what we might today call an hourglass shape. *Source*: Getty Images.

sun with rays that end in hands. Aten is thought not to be male, as Ra-Horakhty was, but to be an androgynous god (Robins 2008: 149). King Amenhotep and Queen Nefertiti "represented the generative force of the male and female principles of the universe that had been separated out from the androgynous creator at the time of creation" (Robins 2008: 149). Perhaps because their god was androgynous, an androgynous look became the ideal for men and women. We see this depicted in the bodies of art from this time, as well as the garments (Figures 10 & 11).

The bodies as depicted in Amarna art are characterized by "long faces and necks, narrow shoulders, short upper torsos, prominent buttocks, large thighs and spindly limbs" (Robins 2008: 148). The effect of this is that the men appear to have round hips, and in some depictions their pectorals resemble breasts, creating an overall effect that today we might call an hourglass shape. Egyptologist Gay Robins writes,

> It is well-known that in both two- and three-dimensional Amarna art, figures of the king are feminized, with wide hips, rounded bellies, high waists, large breasts, narrow shoulders, and slender limbs. Because of this peculiarity, there are cases where the actual sex of the person depicted has been in dispute. (Robins 1993: 29)

Given this rapid change, the fact that the king was in charge of this artwork, and the great quantity of works in this style from the period, we can infer that this body type and style of adornment was thought to be an ideal.

Some fabrics from this time period do remain today (Carter 1922/1977) but they are very fragile, and our most extensive source of evidence of what nobility wore in the time period are in depictions carved into bas relief stone, and other three-dimensional sculptures (Robins 1993, 2008). In these works, we can see that the garments worn emphasize the ideal hourglass figures for both the king and the queen.

In depictions of Akhenaten from the Amarna period the natural line of the body cuts in not at the hip but just below the pectorals, which are rounded at the side like breasts (Robins 1993, 2008). The "feminized" shape of the king is emphasized by the garments he wears (Robins 1993, 2008: 153). In a picture in Robins's book *The Art of Ancient Egypt* we see in an image of Akhenaten that he wears a piece of jewelry at the center of the waist, which draws the eye in here (Robins 2008: 153). The king is wearing what we might call a skirt or kilt. This kilt is rounded at the waist to emphasize the pot-belly shape. The belly button is horizontal, rather than vertical, a departure from the norm (Robins 2008: 153), which suggests softness and a higher percentage of body fat. This round shape is further emphasized with an exaggerated belt, with detailing that draws the eye across this curved belly line. The radiating pleated folds of the kilt serve to make the roundness of the hips more prominent. Unlike modern Scottish kilts, which have pleats that go straight down, and narrow the hips, the kilt on this statue of Akhenaten has pleats or folds which radiate from the center out toward the hips.[1] This causes the hips to appear rounder than they would if the pleats had not been present, or had they been vertical.

While King Akhenaten wears garments that make his body look rounder and softer, Queen Nefertiti, in contrast, is shown in garments that minimize the hips, making her body appear more masculine. We see in a limestone altar, a painted limestone bas relief slab, and paintings in the burial chambers of Tutankhamen that Nefertiti is shown in garments that have a vertical line at the waist (Robins 2008). She is seen in long ties that start at her waist, and fall to mid-calf, as well as in skirts that have vertical stripes or pleats (Robins 2008). Overall, the lines on Nefertiti's garments minimize and flatten the appearance of her hips. She also is sometimes depicted as large as the king (Robins 1993: 37).

The differences between Akhenaten and Nefertiti in the lines of these garments are seen across a number of representations and, given how carefully these artworks were constructed, would not have been an accident. We see in these garments imitation of natural meaning that leads to a more androgynous look. The imitation of natural meaning seen in the garments of Akhenaten and Nefertiti may have been motivated by a desire to lessen the appearance of sex-specific traits, thus bringing Nefertiti and Akhenaten closer to their

Figure 11 Statuette of King Akhenaten and Queen Nefertiti at Tell el-Amarna, 1353–1337 BC. In this statue we see King Akhenaten and Queen Nefertiti depicted in a way that emphasizes his roundness and makes her look more vertical in the hips. Notice how the direction of the pleats and the lines of the fabric draw the eye to different parts of the body. *Source*: Getty Images.

androgynous god. These cases from the Amarna period also show that a desire to emphasize rigid gender binaries is not found in all cultures throughout all times in history.

Imitation of Natural Meaning in Prehistory: Red Ochre

With these cases from Medieval Europe and ancient Egypt we see that our practice of using grooming techniques and garments to imitate natural bodily features is not new. We might wonder if the Egyptian cases are the oldest instances of imitation of natural meaning. How far back in history can we find evidence of this behavior?

Consideration of this question takes us back in history far beyond the ancient Egyptians to a fascinating case from around 100,000 BC. In their 1995 paper "The Human Symbolic Revolution: A Darwinian Account," Knight, Power, and Watts discuss cases of red ochre that were found at a dozen sites (Knight et al. 1995: 85). Red ochre is a substance made of a red rock that is soft enough to be ground into a fine powder that is mixed with water to create a paste and is then used as a pigment. Even today ochre is still used by artists as a way to "sketch" out their paintings on canvas. I personally have some red and yellow ochre-colored paint with my set of oils at home. Red ochre is also the substance that was used on a number of famous cave paintings, such as Lascaux.

Red ochre is not used just as a pigment in paintings but also as a way to color the body. It has been found at numerous burial sites around the world (Knight et al. 1995; Hovers et al. 2003). There are a number of possible functional purposes the ochre could have served. It could be used as a technical aid in tanning or hafting, as sunscreen, to keep off pests, and so on (Hovers et al. 2003: 505). In the archaeological literature there is an implied dichotomy—sometimes made explicit—between interpreting the ochre found at some site in terms of these functional purposes, and the ochre being, instead, symbolic (Knight et al. 1995; Hovers et al. 2003).

Knight et al. argue that the ochre was symbolic. Like Barthes, they use structuralism as their symbolic framework, and make this quite explicit writing, "Our 'time-resistant' mythico-ritual syntax owes much to Levi-Straussian structuralism" (Knight et al. 1995: 103). Indeed, there is a long history of archaeologists and anthropologists using structuralist theories in their interpretive work (Conkey 2001; Trigger 2006; Johnson 2017).

Structuralism has a sort of theoretical clarity that makes it helpful to theorists who discover artifacts and aim to interpret their meaning. Structuralism as it has been applied by anthropologists also draws on a dichotomy between opposites: day and night, masculine and feminine, light and dark, sun and moon, and so on (Leroi-Gourhan 1968; Knight et al. 1995; Trigger 2006; Johnson 2017). This framework is limited but it is easy to see why it would appeal to anthropologists and archaeologists; if there is any plausible theory of meaning that is universal for all of humankind, perhaps it is these opposites.

In interpreting archaeological evidence such as ochre and shell beads, some philosophical framework is needed. This is either recognized and made explicit by the authors or implied by their interpretation of the evidence. No interpretation is done without some implicit or explicit theory about what the evidence means, and why, in archaeology or science more broadly. Discussions of prehistoric bodily adornment as a system of communication explicitly or implicitly rely on philosophical machinery about how artifacts can be bearers of meaning. We have seen how Roland Barthes drew on semiotic theory in his analysis of fashion. We find the same thing in archaeology and anthropology. This started in the 1960s and continues through today (Weissner 1983; Conkey 2001; Hovers et al. 2003; Trigger 2006; Preucel 2010; Bauer 2013; Crossland 2014; Johnson 2017).

In some discussions of bodily adornment a theory of communication is not explicitly detailed but is implied by (1) the sort of evidence that is taken to demonstrate symbolic behavior, and (2) the sorts of messages that are proposed as the communicated content (Knight et al. 1995; Kuhn and Stiner 2007; d'Errico et al. 2009; Stiner 2014; Hiscock 2014). Within archaeology and anthropology the practice of drawing explicitly on semiotic theory is called "structuralism," and is characterized by "a body of ideas about how human culture and the human mind work" with "explicit origins in linguistics and the study of language" (Conkey 2001: 274). Within structuralist archaeology theorists assume that any object of interpretation, from red ochre, to shell beads, to cave art "was generated from a set of underlying cultural premises that are structured like language" (Conkey 2001: 274). It is clear that Roland Barthes was not the only one to find potential in semiotics for interpreting human behavior.

In his analysis Roland Barthes drew on Saussurean semiotic theory, and Saussure's notion of signifiers and signifieds. This tradition of semiotics is sometimes used by archaeologists as well (Leroi-Gourhan 1968; Conkey 2001; Trigger 2006). The most explicit and thorough case of the Saussurean semiotic theory being applied to archaeology was done by Andre Leroi-Gourhan. Similar to Barthes's systematic analysis of French fashion magazines from the 1950s, Leroi-Gourhan undertook an exhaustive analysis of caves. He published this work in an enormous 1968 book consisting of 543 pages, 56 diagrams, and 739 photos—110 of which are beautifully printed in color. Leroi-Gourhan understood the world in terms of binaries of day/night, masculine/feminine, warm/cold, and so on. He interpreted the caves themselves, as well as the things found therein within this schema.

In his appendices at the back of the book Leroi-Gourhan classifies the drawings in the caves themselves as masculine or feminine. He wrote of the famous French cave Lascaux, "The cave as a whole does seem to have had a female symbolic character, which would explain the care with which narrow passages, oval-shaped areas, clefts, and the smaller cavities are marked in red, even sometimes painted entirely in red" (Leroi-Gourhan 1968: 174). Leroi-Gourhan's theory ran into problems when he used it to make predictions about what things would be found in future caves. This took his theory out of the realm of the theoretical and made it falsifiable. And with these predictions—which did not hold—Leroi-Gourhan thereby falsified his theory (Lewis-Williams 2002: 64; Curtis 2007: 161–164; Bahn 2012: 28).

In their interpretation of the red ochre Knight et al. make use of this structuralist framework of opposites, which they present explicitly. In their theorizing they look to "the persistent mythico-ritual linkages between the moon, menstruation and hunting luck" (Knight et al. 1995: 103). It is here that Knight et al. introduce what I take to be prehistoric imitation of natural meaning: they propose that red ochre as used to simulate menstruation, which they take to be an indicator of "impending fertility" (Knight et al. 1995). That is, on their view women would engage in the deceptive practice of smearing their inner thighs with red ochre to select for those men who are "prepared to wait around" (Knight et al. 1995). Their theory is that this practice would lead to fathers who will eventually be better providers for offspring, because the capacity to wait is a sign of a personality trait that would be beneficial to raising offspring.

Knight, Power, and Watts argue that this practice of using red ochre to simulate menstruation began around 105 kya and again resurged around 40 kya in the transition from the Middle to Upper Paleolithic (86–7). Their argument for this is that during a time they date as roughly 105 kya "copious amounts of [red] ochre are ubiquitous in cave/rockshelter sites" and that the data "pass the 95 per cent confidence limit that they [the red ochre findings] are not attributable to sampling variation alone" (87). Knight et al. spend some time rejecting a functional explanation of the ochre, concluding that "more utilitarian uses of iron oxide were secondary to ritual body painting" (104). In summary, their position is that the presence of red ochre—which does not appear to be random or to have served a primarily functional purpose—was used by women to fake menstruation, thus providing an advantage to their offspring, a practice that began 105,000 years ago and again resurfaced around 40,000 years ago.

Now, there is reason to be skeptical of the Knight et al. interpretation of the red ochre. There are many points in their argument to which one could object. Would menstruation really be viewed in this way? Would the women choose to fake menstruation in this way? Would the men really be fooled? Would this really occur across cultures and time in this way? Have all other alternative explanations been sufficiently ruled out? It would be reasonable to argue that the answer to all of these questions is "no." And if their argument is right, why, then, wouldn't this practice continue today?

But before we reject the Knight et al. proposal outright let us consider it a bit more. First, it is important to note that if this use of red ochre did occur, it would be a case of imitation of natural meaning. If Knight et al. are correct in their hypothesis, then this red ochre was used intentionally, deceptively, and had meaning—but that intention was not intended to be recognized by the interpreter.

We see then that a Gricean explanation (the framework of natural meaning) is demanded to explain the potential red ochre used to simulate menstruation. A premise in Knight et al.'s argument that menstrual simulation occurred is the claim that real menstruation itself has natural meaning. Following from the Knight et al. argument, we could say, of a particular case, echoing Grice's case of the measles,

25) That menstrual blood means that she will soon be able to bear a child.

In other words, real menstrual blood had natural meaning according to Knight et al. When it is fake menstrual blood, simulated by red ochre, (assuming for the sake of argument that this took place) the inference on the part of the viewer would be the same. But, in the case of ochre menstrual simulation it would not be factive. This is because a woman could apply red ochre to her thighs when she is not actually menstruating, and could not actually soon carry a child.

If pigments, such as red ochre, were used to simulate natural features that are themselves taken to have natural meaning, then we have an instance of imitation of natural meaning in prehistory. Now, as I have noted here, Knight et al. have not provided sufficient evidence that this is indeed what was going on at this site.

However, it is an important case because it turns our attention to the possibility of this type of meaning with bodily adornment in prehistory.

Interpreting Prehistory

Are there other cases or possible cases of imitation of natural meaning in prehistory? With this question, as with all questions about prehistory, we must keep in mind some facts about preservation of evidence. Artifacts degrade over time, on a timescale that varies with the physical material itself, and the surrounding conditions (Renfrew and Bahn 2012). A number of prehistoric paintings and artifacts have been found in caves, but this is not because of any special affinity our ancestors had for caves. Rather, this is due to the simple fact that caves provide the right conditions for artifacts to be preserved, and in difficult-to-access caves artifacts can be left undisturbed for thousands of years, and sometimes tens of thousands of years. Also, with cave paintings and other markings on cave walls more generally, the slow flow of minerals mixed with water across their surface, as occurs naturally in caves, serves to seal them in, protecting them from damage (Pike et al. 2012). The mineral flowstone in caves also provides an opportunity to do Uranium-series isotope dating, so we can know how old the sites, paintings, and artifacts are (Pike et al. 2012). Any artifact from prehistory that is available for modern analysis has made this unlikely journey from its creation by early humans and stayed intact in its context long enough for us to discover and interpret it.

Peircean Interpretation in Archaeology: The Case of Qafzeh Cave

The interest in discussing archaeological interpretation in terms of semiotics is pervasive. Where Knight et al. drew on the Saussurean theoretical framework of opposites in terms of signifiers and signifieds, other authors have drawn on a Peircean framework to interpret similar findings. Charles Peirce, an American philosopher, also working around the start of the twentieth century developed his own theory of semiology independently of Saussure (Chandler 2007). Unlike Saussure's twofold distinction between signifier and signified, Peirce draws a threefold distinction between icon, index, and sign (Weissner 1983; Conkey 2001; Trigger 2006; Preucel 2010; Bauer 2013; Crossland 2014; Johnson 2017).

In a 2003 paper Hovers et al. discuss a case of ochre found in Qafzeh Cave, in Israel. The timeline of the evidence they consider is about 100–90 kya (494). "The occurrence of ochre in these contexts [Middle Paleolithic] and its use by hominids of this period are now uncontested. What is still controversial is the behaviors that these remains represent" (504). As we have seen with previous researchers, Hovers et al. again discuss their findings in terms of a dichotomy between practical purposes and symbolic purposes. They write that to conclude that the ochre served a symbolic function, "one would need to demonstrate that ochre was positioned within an organized system of other symbols and thus had referential powers and the capacity to generate symbolic predictions among the participants in the symbolic network" (504). Hovers et al. then rule out to their satisfaction the possibility that the ochre can be explained in terms of a technical aid in tanning or hafting (505).

Functional or Symbolic: A False Dichotomy

Let me make a note here about the interpretive possibilities—a point that applies as much to prehistory as it does to today. Hovers et al. state that coming to the conclusion that some part of the archaeological record has a symbolic function "is only acceptable only after other, testable explanations have failed to fit the archaeological evidence" (506–507). They write that "other explanations for the occurrence and the characteristics of the ochre assemblage are less parsimonious than a symbolic one" (493). Hovers et al. conclude that the use of ochre as found in these caves did not serve a functional purpose and is thus "an early case of color symbolism" (Hovers et al. 2003).

We might question the dichotomy Hovers et al. set up between the ochre serving *either* a practical purpose *or* a symbolic one. We see that in other cases of meaning in bodily adornment something can be both. For instance, cowboy boots serve a functional purpose in that they have a pointy shape that can easily slip into a stirrup, slight heels and sometimes spurs that can be used to steer a horse, and come up past the ankle so that mud and manure don't get inside the shoe and onto the foot. One need only imagine approaching, mounting, riding, and steering a horse in flip-flop sandals to see the functional advantages afforded by cowboy boots.

But a cowboy boot isn't *only* functional—it serves these functions and *also* serves a symbolic function. In the right context you can identify someone who rides horses by their shoes. Sometimes an item of bodily adornment—such as cowboy boots—is taken out of its original context and into another, where it gets an even richer symbolic meaning. This is seen in cowboy boots when they are worn by politicians—incongruously—with suits as a "way to signal regional identity" (Fernandez 2014; Mulkern 2016). Former U.S. representative from Utah, Jim Mateson, has been quoted saying that when he sees someone wearing cowboy boots he thinks "that's someone I'm going to relate to" (Mulkern 2016). Former representative from Colorado John Salazar has joked that even in Washington he needs cowboy boots to avoid stepping in metaphorical manure (Mulkern 2016). If archaeologists in the future were interpreting cowboy boots to *either* be symbolic *or* functional but not *both* this would fall prey to a false dichotomy and miss out on all these many shades of meaning. We cannot assume that the same was not true for ochre in prehistory.

This does not mean that Hovers et al. are wrong that the ochre served a symbolic purpose, but simply that it is not supported by their arguments that the ochre did not serve a functional purpose—nor does it require this. Because red ochre has a striking color we can assume that when used on the body it would have acquired a symbolic meaning, even if it first served a functional purpose as the cowboy boots did. Their conclusion may be right but it is not supported by their argument.

Hovers et al. discuss the symbolic meaning of the red ochre in Peircean terms. Although they don't explicitly state their intention to use the Peircean framework, this is implied by the terms in which the findings are discussed. At different points in the article the authors speak of reference, icons, indexes, symbols, resemblance, and intentions (Hovers et al. 2003). We see in such explanations the Peircean framework in the background.[2] But the authors are not precise in making these attributions of meaning. Red ochre may very well have served a range of symbolic purposes, as proposed by Hovers et al., and previously by Knight et al., but in many of these discussions the claims about what meaning these may have signified reveal some fallacious reasoning.

Non-Natural Meaning in Human Adornment

Another potential case of meaning in prehistoric bodily adornment is shell beads, which are found around the world at a range of times in human history. Will beads prove to be more conclusive as bearers of meaning than red ochre did?

In discussions of shell beads, as in the Hovers discussion of red ochre, we see symbolic communication being heralded as an important step toward modern humanity. In his 2009 paper "Additional Evidence on the Use of Personal Ornaments in the Middle Paleolithic of North Africa" archaeologist Francesco d'Errico and his coauthors present their findings consisting of a vast number of shells from the sites Rhafas, Ifri n'Ammar, Contrebandiers, and Taforalt in modern-day Morocco. To demonstrate that these were artifacts, the authors must demonstrate that the shells were modified and used by humans. Of course, a number of shells that are found in excavations are not understood to be artifacts but simply animal remains. Also, sometimes shells found at sites were brought there by humans, but are thought to be in a specific location because the meat that had originally been inside was a food source. This is the case for shell middens, which are thought to essentially be prehistoric trash heaps (Johnson 2020). Such shells result from intentional behavior, but not intentionally communicative behavior (Johnson 2017). That is, the shells in a midden end up there because of intentional, goal-directed acts, but the shells were never intended to communicate anything. Indeed, the shells themselves were not a part of the goal, but a means of achieving the goal. We all perform countless acts like this each day. For instance, this morning I made coffee by putting grounds into a filter in a coffee maker. This is because I wanted coffee. Those grounds are now in the trash. I put them there intentionally but not to communicate anything.

In his 2009 paper d'Errico begins by arguing that the prehistoric shells found in the North African sites are the result of intentional behavior, and then argues that this was intentionally communicative behavior. In arguing for his first point—that the shells were the result of intentional behavior—he begins with geological evidence. The first relevant question to consider is whether or not these shells could have naturally occurred in the location in which they were found. Essentially the first thing that we need to rule out is that these sites were underwater or places where shells could have ended up through currents or the agency of the shell-creatures themselves. On this point d'Errico writes, "The shells do not derive from the bedrock in which these cave sites are formed." In other words, the shells did not originate in these locations. Rather, "The shells were deposited at these sites, a distance too great for natural processes known to carry marine shells inland" (16053). These shells did not come to this location via natural processes such as flooding events.

D'Errico next rules out the possibility that these shells were part of a midden resulting from the shell-creatures being used as food—the prehistoric equivalent of my coffee grounds in the trash. He writes that this was not the case because "they were collected dead on the shore," and not fished—as shellfish would be for eating. Thus, he concludes that the shells "cannot be interpreted as the remains of subsistence-related activities." In other words, the shells at these sites were brought to their location intentionally, and that intention was not to eat the creatures inside. Lastly, d'Errico concludes that the shells were collected with an eye toward particular species and not a random collection, writing that the findings were "not the result of a random collection of dead shells on a shoreline" (16053). As we have seen in the work of other archaeologists, d'Errico leads us to his conclusion that the shells had symbolic meaning by ruling out other explanations. If we agree with d'Errico's argument thus far, we have now ruled out the possibility that the shells in his study originated at these locations were moved there by non-human forces, were accidental, the result of idiosyncratic behavior, or from the creatures inside being used as a source of food.

Of course, ruling out explanations is easier than supporting your own hypothesis, but it is to this task that d'Errico turns next, and with his use of imaging techniques, rather successfully. D'Errico argues that these shell beads were worn on the body. If this can be demonstrated it is difficult to argue against the point that these were items of bodily adornment. With the red ochre there is always the possibility that the ochre served some practical purpose, such as keeping off the sun or insects, but it is difficult to imagine what a practical purpose of wearing shell beads would be.

The way that d'Errico argues for the shell beads having been worn is rather ingenious. What d'Errico has done is conduct microscopic analysis of the shells looking for the sort of micro-scratches that would result from the beads rubbing against fabric. "Direct evidence for the use of the shells as beads comes from the identification of eighteen *Nassarius* from Taforalt and the five shells from Rhafas of a use-wear pattern different from that recorded" (16054). He continues, "Two other types of wear patterns, absent in natural collections and likely related to the stringing of shells, are recorded on most of the discovered specimens" (16054). In addition, d'Errico performed microscopic imaging analysis to confirm that the holes in the beads were made by hand and not through natural causes. This is because there are tiny chips and scratches around the sides of the holes that demonstrate a tool was used to make them. So, we have a strong argument supported by the empirical data that backs the hypothesis that the shells were intentionally moved, modified, and worn.

Further, there is evidence that the shells were made to be different colors, either through the application of pigment, which remains on some of the samples, or through the shells having been heated on a fire which causes them to turn black in color (d'Errico et al. 2009). D'Errico and his team meticulously photograph this evidence. On examining the photographs and reading the arguments there seems to be little doubt as to the merits of their interpretations of the markings on the shells. Further, the striking opalescence of the shells revealed in the photographs makes it easy to imagine their aesthetic appeal in 70,000 BC.

In discussing these findings, d'Errico writes, "Symbolic material culture, representing the ability to share and transmit coded information within and across groups, is an indication of modern cognition" (d'Errico et al. 2009: 16051). Here d'Errico focuses not on the capacity for symbols themselves but on the capacity to create cultural items that convey symbolic meaning. He writes that "Beadwork represents a technology specific to humans used to convey social information to other individuals through a shared symbolic language" (d'Errico et al. 2009: 16051). Although later in the Chapter I will question whether humans were the only creatures to have complex symbol systems, certainly it cannot be debated that humans are the only ones to create and use beads to convey information.

Although d'Errico has supported his argument with a plethora of systematically researched empirical premises, no scientific instrument can measure meaning. At the end of the discussion of the empirical evidence d'Errico concludes that "Analysis of the shells from Rhafas, Ifri n'Ammar, Contrebandiers, and the specimens from Taforalt reinforces the conclusion that they were transported to these sites by humans during the Aterian and used for symbolic purposes" (d'Errico et al. 2009: 16053). Although we have evidence that the beads were worn, it remains to be proven that these were used for symbolic purposes. D'Errico does not take the steps necessary to lead us to this conclusion, unlike with the rest of the careful steps in his paper. Of the empirical points he writes, "These observations reinforce, rather than weaken, the argument for their symbolic use" (d'Errico et al. 2009: 16053). For his evidence of symbolic use he instead provides only this gesture toward a proposal, writing, "Ornaments made of slightly modified natural objects often represent, by the direct link they establish between the natural world and the meaning attributed to them, quintessential symbolic items" (d'Errico et al. 2009: 16055). It is not clear what data d'Errico has in mind for this claim that "slightly modified natural objects often represent" something "by the direct link they established between the natural world and the meaning attributed to them." In his focus on the "direct link" between the "natural world" and the "meaning" d'Errico seems to be suggesting that shells beads acted as a sort of index, in that they come from the sea and thus indicate some meaning *about* the sea.

It is not a leap to imagine wearing objects from the sea as adornment because we still do this today. Pearls are perhaps the most commonly worn sea-adornment in the West. In 2018 annual pearl sales in the United States totaled an estimated 1.3 billion dollars, and by 2022 are predicted to total 16.8 billion dollars globally (Danziger 2018). Pearls come from the sea. But—crucially for this discussion— the wearing of pearls does not mean anything about the ocean. Instead they evoke a sort of staid WASP-i-ness—hence the expression "clutching her pearls." Modern jewelers are trying to overcome their connotations as something worn by older women and report that customers want to "avoid the 'mother of the bride' look" (Danziger 2018). It seems plausible, as d'Errico suggests, that the shell beads did serve a symbolic purpose. But, again drawing on what we know about how meaning with bodily adornment works for us today, the evidence is not sufficient to demonstrate that shell beads indexically signified something about the sea.

The Case of Shell Beads: Stiner

Let me now consider my final prehistoric case study before moving on to consider the possibility of animal adornment in the remainder of the chapter. In her work, archaeologist Mary Stiner highlights the complexities involved in attributing specific meanings or types of meanings to items of prehistoric bodily adornment. In articles from 2004 and 2014, Stiner discusses beads from a number of sites around the world. Although she understands beads as "instruments of symbolic communication" (Stiner 2014: 51), she remains circumspect about exactly what they mean. Instead she focuses on the formal properties of the beads themselves and discusses the ways that the structure of beads could facilitate complex meaning. Stiner draws attention to the fact that beads are durable, standardized, quantifiable, transferrable, costly, and may be compounded and recombined. She argues that these features support the hypothesis that beads are the sorts of entities that could be used as signs (Stiner 2014; Kuhn and Stiner 2007).

Notably for our discussion, the features of beads that Stiner draws attention to are important from a philosophy of language perspective. In linguistics and philosophy there are a number of varying accounts of exactly what features something must have to count as a *true* language (Von Frisch 1974; McWhorter 2003; Senghas 2004; Pietrosky 2018). For instance, in her research, linguist Ann Senghas argues that Nicaraguan Sign Language "counts" as a *true* language and not "mere" gesture because it has the properties of discreteness and combinatorial patterning (Senghas 2004). Nicaraguan Sign

Language is a language that was invented by deaf Nicaraguan school children when they were put together in classrooms for the first time in the 1970s and 1980s. As Senghas documents in her work, Nicaraguan Sign Language has discrete units that can be recombined, and their meaning depends on and changes based on the positioning within the greater whole (Senghas 2004). Senghas argues that because Nicaraguan Sign Language has these particular features and structures it is importantly different from gesture. Gesture as she understands it does not have parts that are recombined in meaningful ways (Senghas 2004).

Stiner is right to emphasize that prehistoric beads are, as she says, durable, standardized, quantifiable, transferrable, costly, and may be compounded and recombined. Because of this, prehistoric beads bear the hallmarks of language. And—crucially for the discussion in this chapter—red ochre does not have these properties. Red ochre is just one color, where the beads are found in their original white and opalescent shades, as well as black (burned in a fire) and red (colored with ochre) (d'Errico et al. 2009) which presents the possibility of a multiplicity of meanings that could be represented through color-specific patterns. Red ochre is a blunter instrument for use within a possible system of communication. It is possible that the ochre was painted on the body in careful patterns such as dots and stripes, but its lack of durability presents natural limitations to its possible use as a conveyer of meaning—limitations that are not present with the shell beads.

Although Stiner spends time drawing attention to the formal properties of the beads, she does not give details on what she thinks beads symbolized. As might be clear from my discussion earlier, I take this to be a strength—not a weakness—of her argument. Stiner writes that although beads "seem to have been the minimal particles of information, loosely akin to bits in a digital coding system" there is "little to be gained from trying to reconstruct the actual messages conveyed" (Stiner 2014: 62) She focuses on the likelihood that the meanings of beads "were so free to vary" (Stiner 2014: 62). This means that for Stiner the beads are not icons or indexes, but rather have the potential to be signifiers or symbols with arbitrary, language-like relationships to the referents.

There is a vast difference between the sorts of conclusions that can be drawn from the archaeological record when signs are understood as necessarily arbitrary versus as indexical or iconic. As we saw with pearls meaning is unpredictable and doesn't necessarily bear a clear indexical or iconic connection to the source. The Stiner position is the most measured and principled discussion of the meaning of prehistoric bodily adornment that we have seen. She does not present a false dichotomy between function and meaning, and she wisely says that the likely meaning of the shell beads is arbitrary, and language-like.

Uniquely Human?

In introducing her work on the ochre finds of the Qafzeh caves Hovers wrote, "For many researchers the ability to create arbitrary relationships between ideas and their referents—that is, to construct and use complex symbol systems—is the defining characteristic of Homo Sapiens" (491). Is this claim true? Let us now consider whether or not our behaviors of adorning and using symbols are unique. There are a number of terms in this sentence that are worth a close look, and the truth or falsity of this claim depends on exactly how we understand them. Let's start with "complex symbol systems."

A number of animals have systems of communication involving their bodies that are complex. Let us consider again the dance of the bee. As discovered by Karl von Frisch in his Nobel Prize–winning research the dance of the bee is remarkable for a number of reasons. The information conveyed is complex and it is transmitted with incredible accuracy, as evidenced by how the information is acted upon by the receivers of the message. When I teach von Frisch I have undergraduates hide candy in the building and then draw pictorial maps for a partner depicting its location. I ask that their maps contain no writing or other symbols. With a few exceptions, they are amusingly unsuccessful in this task. Students end up wondering the halls with nothing more than a few vague lines or squiggles on a page to guide them. But this is the point. Getting someone to a specific location and a specific distance away requires a great deal of information. Somehow the bees have figured this out, and the survival of the species depends on it.

So, then, we might wonder if the bee dance is a counterexample to the Hovers claim that "the ability to create arbitrary relationships between ideas and their referents—that is, to construct and use complex symbol systems—is the defining characteristic of Homo Sapiens" (491). It cannot be denied that the bee dance is complex, on any understanding of the term. But the bee dance still may not constitute a counterexample, because of Hovers's inclusion of the arbitrariness condition. She writes that the defining characteristic of *Homo sapiens* is creating *arbitrary* relationships between ideas and referents. What is meant by arbitrary here and why would Hovers include it? For this answer we must return to the Peircean framework.

Recall that Peirce introduces a threefold distinction between icon, index, and symbol (Peirce 1998). Something refers to something else iconically when there is some sort of resemblance relation between the means of referring and the thing referred to. For instance, on a map, I can tell that it will take me longer to drive from San Diego to Yosemite than to Sequoia because the distance on the map is greater. Resemblance is a tricky notion[3] but in some way the differences between these distances on the map resemble the difference in the time it will take me to drive. As with the bees, there is a very precise correlation between the duration of the bee dance and the distance the bees must fly to reach the food. This means that there is a resemblance relation underpinning this act of reference. In other words, the sign indicating the distance to the food source is not arbitrary. There are specific features of the sign itself (the dance) that correspond to the thing being referred to (the distance to the food), thus making the length of the dance an *icon* in Peircean terms. Although I won't walk us through it here, we can see that there is also this sort of resemblance relationship between the angle the bee needs to fly to reach the food, and the angle at which the bee dances on the wall (Von Frisch 1974). That means that this information about direction, too, is conveyed *iconically* as well.

Another piece of information that is conveyed to the interpreter bees is the type of pollen and the sweetness of the nectar. This is communicated through scent (Von Frisch 1974) and is the result of a causal relation between the sweetness, type of pollen, and the way the bee smells when she comes back from the food source. There is a causal relationship between the sign and the referent, and thus the scent of the pollen acts as *index* of the type and sweetness of the nectar. Scent indices are a very reliable sign, for the bees and for us. One need only think of how the smell of cigarette smoke, alcohol on the breath, and so on are interpreted by us to understand how this could work. We know these scents are caused by certain behaviors and take their presence as solid evidence of these behaviors having taken place.

So, we see in the bee dance that we have icons, and an index. Do we have, too, a symbol? Indeed we do not, and it is in this fact, and in Hovers including the arbitrariness clause that we can preserve the notion that the defining characteristic of *Homo sapiens* is the ability to create and communicate with arbitrary signals is preserved. We know that animals use icons and indices. In seeking to identify uniquely human communicative behaviors within a Peircean framework the focus should be on identifying *symbols*, that is, on identifying things that bear an arbitrary relationship to their referent.

Is Adornment Uniquely Human?

Considering bodily adornment is vitally important to understanding humans as meaning-makers. The practice of adornment in our species operates in a number of different ways, as we have seen. But is the practice of adornment unique to our species? This is ultimately a philosophical, metaphysical question.

As with so many other questions in philosophy the answer hinges on exactly how we understand "adornment." We have seen that a number of creatures, such as Darwin's case of the peacock, have bodily features that do not serve some functional purpose and are there because of the forces of sexual selection. More broadly, as discussed throughout the book, animals certainly have coloring, patterning, feathers, changes in expression, changes in bodily features and so on, that communicate information. But any individual peacock has no control over what his feathers are like. So we see that we can draw a distinction here between possible cases of adornment that are bodies themselves versus as a result of behaviors or due to emotion.

Cephalopod Color Change

As Darwin noted, animals change their bodily posture when they are in different emotional states. He commented on a dog holding his ears differently when he is happy or sad. Posture can be moved easily and undoubtedly some of this is unconscious in animals as it is in us. We too change our posture if we are excited or sad, but this is not an intentional, conscious change.

Cephalopods such as cuttlefish and octopuses can manipulate their body shapes in a number of ways due to being invertebrates. They can squeeze through small holes which makes them difficult to keep in captivity (Godfrey-Smith 2016). They also can change the color of their skin. Philosopher of biology Peter Godfrey-Smith, working with biologist David Scheel, has observed octopuses changing colors and has tracked which behaviors correspond to which colors (Godfrey-Smith 2016; Scheel et al. 2016). Usually octopuses are thought to be solitary but at a rare site in Australia they found a number of octopuses living in close proximity and named it "Octopolis." According to their observations, some of the color changes seemed to be random, the equivalent of "chatter" not linked to observable behaviors (Godfrey-Smith 2016: 184). But Scheel and Godfrey-Smith also observed that octopuses turn dark as a marker of aggression (Godfrey-Smith 2016; Scheel et al. 2016). Godfrey-Smith writes, "When an aggressive male is about to attack another octopus, he will often turn dark, rise out of the seabed, and stretch his arms out in a way that magnifies his apparent size" (Godfrey-Smith 2016: 184). They call this the "Nosferatu pose" after the silent film vampire. This apparent increase in size is an instance of imitation of natural meaning, the underwater equivalent of a bear standing on its hind legs, with its hair sticking up on end. Not only is darkness an indicator of aggression but it can also express the *degree* of emotion. The darkness is a reliable predictor of how aggressive the octopus will be (Godfrey-Smith 2016: 185). In contrast, pale displays are seen "when an octopus is not willing to fight" (Godfrey-Smith 2016: 185). For further future research toward deciding whether or not this is adornment we would want to know the degree to which this is within the control of the octopus, or if it is more like our automatic behaviors of blushing or having our arm hair raise. Is it something the animal *does* or something that *happens* to the animal?

Wallowing

What about potential cases of adornment that are not a result of genetic inheritance of bodies or emotional response but a result of behaviors? I have already discussed bird feathers a great deal, and this was a favorite topic of Darwin's. But I have not yet discussed behaviors of birds or other animals that resemble our adorning behaviors of applying makeup or ochre to the body.

Humans are not the only living species that covers the body in substances. A number of other animals engage in wallowing—rolling in mud—which covers their body in the mud substance. Interpreting wallowing behavior is difficult. One possible functional explanation of wallowing behavior in wild boars is that it is done to serve some functional aim such as removing parasites from the body (Fernandez-Llario 2005: 9). Other possible explanations for this behavior are that it is done to thermoregulate—to cool off—or to disinfect wounds that were sustained in fights (Fernandez-Llario 2005). In a 2005 study biologist Pedro Fernandez-Llario challenged the idea that we could explain the wallowing behavior of wild boar solely in these functional terms and argued instead that his findings rule out these purposes (Fernandez-Llario 2005). According to his study, wallowing in wild boars (*Sus scrofa*) is done for the sexual/aesthetic purpose of attracting a mate (Fernandez-Llario 2005). Note that this behavior and the proposed explanations are quite similar to what we saw with explaining ochre use in our own species in prehistory.

Wild boars are not the only species to engage in wallowing, nor is the practice restricted to mammals (van Overveld et al. 2017). Researcher Thijs van Overveld and colleagues studied the wallowing or "mud bathing" practices of Egyptian vultures and argued that they are part of a practice of "cosmetic coloration" used for "social communication" (van Overveld et al. 2017). In the article describing their findings the researchers write that "Most plumage colors are static traits with a relatively fixed information content" (van Overveld et al. 2017: 2216). As we have seen, feather colors are not entirely static because in some species, such as house finches, the red or yellow color of the male finch's head is directly related to the carotenoids found in its diet (Sundstrom 2017). But this is a slow, internal process and quite different from the behavior van Overveld et al. describe as "cosmetic coloration" and "feather painting" (van Overveld et al. 2017: 2216). As the authors note, sometimes birds of other species such as cranes and ptarmigans use soil on the body as a form of camouflage (van Overveld et al. 2017: 2216). The feather painting they observe is significant because it is used in a "social signaling context" (van Overveld et al. 2017: 2216). They write, "At this point, however, it is unclear what other social content, if any, may be signaled through feather painting, especially because we still know very little about Egyptian vulture group dynamics and patterns of social relationship within their society" (van Overveld et al. 2017: 2217). The researchers note that use of the "feather paint" was not restricted by sex or social status, and thus ruled out these possible meanings as what was signaled.

In addition to soil used to color feathers, other birds use secretions from glands to color their feathers (Delhey 2007; Bartels 2017). Secretions from the uropygial gland serve to make the feathers glossier and sometimes these secretions are even tinted (Delhey 2007; Bartels 2017; Amat et al. 2011). In some species the secretion coloration changes when breeding, pointing to a sexual signaling purpose (Delhey 2007). Carotenoids—also the cause of the red or yellow heads of house finches—are found in uropygial gland secretions (Amat et al. 2011). Researchers have shown that flamingoes (*Phoenicopterus roseus*) use the tinted secretions to make their feathers more pink. The authors of this study on flamingoes conclude that "the primary function of cosmetic coloration is mate choice" (Amat et al. 2011: 665). Because application of the tinted secretion is done more when displaying in groups than during the rest of the year, this suggests that the behavior constitutes "deliberate staining" for "cosmetic purposes" (Amat et al. 2011). These researchers do not shy away from describing this behavior in a way that certainly makes it sound like adornment—in this case quite akin to the use of blush that Darwin himself considered.

Because plumage color changes with carotenoids that are eaten, the color of the bird might be taken—as we have seen with the red house finch and flamingoes—to communicate the rich diet of the bird, and thus communicate success in foraging. As the flamingo researchers note, "because plumage colour changes at moult, several months before mate choice, the information conveyed by colour may be out of date when individuals make assessments" (Amat et al. 2011: 666). The authors characterize this signal sent by the application of the carotenoid-rich secretions as "honest" in the sense that the secretions take time and energy to produce and apply (Amat et al. 2011). At the end of the article the authors note that "cosmetic colours may play an important role in courtship and mate choice, and would update the signal value of plumage colour, mainly by providing a more recent snapshot of the bearer's quality than colours acquired by moult some months before" (Amat et al. 2011: 671). However, this assumes that the more recent information is more valuable to the females. It could be that this is a *deceptive* or *imitative* case of natural meaning—something that makes it look as though the bird feathers have this coloration themselves, when in fact it is "merely" the more recent colored secretions on top of them.

One other, final, case of possible adornment by birds that I will discuss is presented in an article by Jared Diamond on bowerbirds. In his work on bower-creating activities Diamond notes that one species of bowerbird, the Fawn-breasted Bowerbird, (*Chlamydera cerviniventris*) holds a "sprig of green berries in his bill" during his courtship display (Diamond 1982: 101). This is different from the bower-making behavior Diamond describes in the remainder of the article because this behavior is an instance of adornment of the *body* rather than of the bower space. As far as the possible meaning Diamond speculates about this, writing that the holding of a green sprig during courtship is a result of this particular species of bowerbird not having a brightly colored crest. With the holding of the sprig the bird imitates a crest "as if to show the female a crest that his ancestors lost" (Diamond 1982: 101). Of course, we cannot know that this is the true aim of the bowerbird (it would require a very complex intention), but it certainly is intriguing, as is so much of bowerbird behavior.

With these cases of possible bird adornment we are left in a position of only being able to speculate as to their meaning. Perhaps we have here cases of imitation of natural meaning. Perhaps we have honest signaling. Or perhaps a bird wishes to imitate a crest that its ancestors have lost (although this is unlikely). What we do see here are cases of birds taking some material from their natural environment and adding it to the body. In some cases they do this more frequently during certain periods than others, as with the application of secretions from glands when in groups (Amat et al. 2011: 665). This at least suggests that there is specific information that is conveyed, although we are left to infer what this might be by observing behaviors. As with the cases from prehistory we are left to speculate in the face of indirect evidence.

With the birds, however, unlike with our prehistoric ancestors, there is the possibility for direct observation of behavior and for future experiments that can provide more evidence in favor of one meaning or another. With his studies of long-tailed widowbirds Malte Andersson has already shown what successful research of this sort would look like (Andersson 1982). Perhaps in the future there could be similar studies done with varying the pinkness of flamingoes and seeing how mating choices change, and so on. It also could be fruitful to research further this distinction raised by Amat et al. between adornment of bodies that express long-term versus short-term states.

Neanderthal Adornment?

The reason that it seems unlikely that a bowerbird intends to imitate a prehistoric crest is because this would require a very complex mental state on the part of the bird. It would require knowledge of history and the ability to think about the minds of others. As we saw previously in my discussion of deception, some birds, such as the piping plover, do have the ability to track gaze, and alter their behavior on the basis of this (Ristau 1990). But as far as we know, for all currently living creatures fully fledged mindreading is restricted to humans (Cheney and Seyfarth 2007). Recall that mindreading is a necessary component of Gricean non-natural meaning: you cannot have an intention that somebody else come to believe something if you are not able to have a belief about their mental state.

The fact that we are the only currently living creatures with mindreading capacities does not rule out the possibility that at some point in history another species or subspecies *had* this capacity (Davies 2020: 72–78). And thus we will end the chapter with consideration of the adornment of *Homo neanderthalensis*. Again, as we saw with the ochre and shell evidence, we must keep in mind the important facts about degradation of evidence over time. When we discuss Neanderthals we are discussing evidence that is older than 40,000 BC, when *Homo sapiens* went into Europe and Neanderthals went "extinct." I put extinct in quotes because Neanderthal DNA lives on in all of us, making up an average of 0.3 to 2 percent of our DNA (Price 2020). Because we interbred there is debate about whether Neanderthals should be considered another species or a subspecies of some broader species category we both belong to (Zilhão 2014; Davies 2020: 76).

Neanderthals undoubtedly engaged in adorning practices that closely resemble ours. They tanned hides to make leather using a variety of techniques (Wragg Sykes 2020: 260). There is evidence that they used ochre for coloring items of adornment including shells (Joyce 2010; Zilhão et al. 2010; Wragg Sykes 2020: 260).

Archaeologist João Zilhão has even proposed the use of what he calls "Glitter makeup" by Neanderthals (Joyce 2010)—reflective bits of shells or other minerals that were chipped off or ground and then used in other pigment mixtures (Zilhão et al. 2010). This might seem surprising. As d'Errico writes, "The manufacture of personal ornaments and bone tools by Neanderthals has been a controversial issue" with staunch defenders on both sides (d'Errico 2003: 195; see also d'Errico et al. 2003, Wynn 2004). Such findings go against earlier hypotheses about the cognitive capacities of Neanderthals such as posited in the 1980s by archaeologist Lewis Binford, who thought there was a wide gap between *Homo sapiens* and *Homo neandethalensis* (Wynn 2004).

In recent years there is mounting evidence that shows the tide has turned on the Neanderthal intelligence debate. Today, some are sweeping in their interpretation of the evidence relevant to Neanderthal adornment. In her 2020 book *Kindred: Neanderthal Live, Love, Death and Art* archaeologist Rebecca Wragg-Sykes writes, "Whatever Neanderthals wore, whether pigment on skin, gleaming tanned leather, cosy furs or threaded red shells, it was always about more than function" (Wragg Sykes 2020: 261). Wragg-Sykes goes so far as to follow d'Errico in proposing an *indexical* relationship between the adornment and the meaning being proposed, writing "Placing things on bodies is a powerful way to express status and identity, and observable to plenty of animals. . . . For Neanderthals, wearing clothing made from hides and furs must have mentally recalled the animals whose bodies they came from" (Wragg Sykes 2020: 261). As I have argued already, we should be skeptical about accounts that propose specific meanings in prehistory. As I demonstrated with the case of the pearls, adornment—even that from a recognizable source—doesn't always mean something about the source. Further, we ought not to posit a false dichotomy between function and meaning. As I showed with the cowboy boot example, we can, and often do, have both. We should take these facts to hold for Neanderthal adornment as well. And although we ought to be skeptical with meanings in specific cases, Neanderthals did undoubtedly adorn the body.

We have seen here the difficulty of interpreting behaviors as evidence for certain types of minds or mental states. We have seen this difficulty again and again with animal minds. We can see a male bird with beautiful feathers act in a certain way. We can see a female bird select a male bird that has beautiful feathers. We can see this play out right before our eyes. But positing mental states on the basis of these observations is tricky. As I will argue in the final chapter, we cannot thus assume that the birds are making judgments of beauty. Interpreting archaeological evidence carries with it these complications, and the additional complication that we are not directly observing behaviors but remains that we take to be evidence of certain behaviors (Johnson 2017; Johnson 2020). We then can try to posit mental states to explain the behavior that would have led to that artifact (Renfrew 1994; D'Errico 2003; Wynn 2004).

Can we say, then, that the sort of adornment we find in Neanderthal sites is an instance of non-natural meaning? We can see that, as with Stiner's beads, we have the right sort of object to carry non-natural meaning. However, this itself is not enough to posit presence of non-natural meaning. This is because non-natural meaning requires the ability for mindreading. If, in the future, scientists are confident that Neanderthals had the capacity for mindreading, then we can see this evidence as a likely case of non-natural meaning in Neanderthal bodily adornment. But without this corroborating evidence the items of adornment themselves are not enough to posit this specific type of meaning—and to claim all this would entail about the mind.

The Significance of Meaning in Prehistoric Bodily Adornment

Let me close the chapter with a final comment on the significance of meaning in prehistoric bodily adornment. There are two importantly distinct types of artifacts that provide evidence for two types of cognitive developments in humans. The first are tools such as handaxes and pottery. These are evidence of human's ability to modify an object to suit a purpose—certainly a large cognitive leap, and one that is shared in the most rudimentary way by only a few other species (Goodall 2010). This ability is clearly a very different sort of cognitive capacity than the ability to intentionally signal to or communicate (Cheney and Seyfarth 2007). As the development of the communicative capacity surely began with spoken word or gesture, rather than in any physical form, as with handaxes or pots, evidence of the development of communication is harder to come by.

Study of items of bodily adornment is of paramount importance to understanding the origins of human communication. As Francesco d'Errico writes,

> Symbolic material culture, representing the ability to share and transmit coded information within and across groups, is an indication of modern cognition. This statement is particularly true when the physical body is used as a means of display. Beadwork represents a technology specific to humans used to convey social information to other individuals through a shared symbolic language. (d'Errico et al. 2009: 16051)

This research on bodily adornment in relation to human language development demonstrates that interest in bodily adornment should not be considered a fringe issue, but, rather, one that can help illuminate the foundations of our systems of communication, and indeed our very humanity.

8

Emotions: Information, Misperception, Suppression, and Expression

Bodies and bodily adornments have been molded by the forces of natural and sexual selection. We have seen numerous instances of this in peacocks, long-tailed widowbirds, bowerbirds, human beings, and other species. In addition to our physical selves having more or less permanent traits such as tails, feathers, nails, and sexual organs, our bodies also reveal passing information about our emotions, both to others and to ourselves. With the invention of adornment such as red ochre and shell beads we developed the ability to express a range of varied changeable meanings. Bodily emotional expression is another important means by which bodies have meaning and will be the focus of this chapter.

As we have seen, Darwin published his theory of natural selection in *The Origin* in 1859 and his theory of sexual selection twelve years later in *The Descent* in 1871. In the interim Darwin published detailed works on orchids, domesticated animals, and other similar, more narrowly focused subjects. In 1872, the year following the publication of *The Descent*, Darwin published another work where again he considers application of his theories to humans—a topic he had avoided in *The Origin*, save for one mention at the end. In *The Expression of the Emotions in Man and Animals* Darwin stressed the links between how we and our non-human animal relatives unconsciously signal our emotional states. For instance, as I have noted, Darwin observed how a dog that is happy will hold his ears in a different way from when he is sad. As another example, Darwin discusses our behavior of leaning in the direction we would like a pool ball to go. Someone may or may not realize they are leaning in the direction they want a ball to go. Our emotional expression is in many ways outside our control. This is in contrast to most cases of bodily adornment. Usually, as adults we can choose what we want to wear, how we want to express ourselves, and can tailor this to an audience and occasion. With emotional expression, rather, we *reveal* our emotional states; we do not consciously divulge them (see Rosenthal 2006; Johnson 2019).

Additionally, cultural factors shape how we show our emotions. It is thought rude to honestly convey certain emotions. For instance, in the West, one tries to not show displeasure when opening a disappointing gift. This is something children have to learn. Women walking in the street are sometimes told to smile, as though the fact that they may be thinking about something other than appearing pleasant is an affront to those they pass. At the same time, certain emotional states can be very difficult to suppress. Someone might attempt to fight back tears while wishing to seem strong, or might try to appear calm while their face is growing increasingly red. Because of this, emotional expression is unlike most ways we intentionally communicate with words. In this chapter I will begin by discussing these forms of emotional expression, which I call "visible emotional expression."

Further complicating the matter of emotions and expression is the fact that the mechanisms underlying the relationship between our minds and our bodies are not as straightforward as one may think. It is not simply that our minds feel a certain way and this is revealed through the body, as illustrated with the examples of tears of sadness or a face growing red in anger. Our bodies send information to our brain as well, which creates a feedback loop. For instance, it has been shown that when participants in a study received undetectable subcutaneous Botox rather than a placebo between their eyes—where the brow furrows in concentration and anger—this resulted in them feeling happier (Finzi 2013). Similarly, taking a painkiller can dull sensations of emotional, psychological pain, such as the pain of rejection (DeWall et al. 2010). Taking a ginger pill can make us judge certain repulsive actions to be less morally objectionable (Tracy 2020). And certain emotional states, such as depression, have a striking relationship to the bodily component of our gut bacteria (Jacobs 2019; Pennisi 2019; Valles-Colomer et al. 2019).

These cases demonstrate that our bodies convey meaning not just to others but to our own minds as well. In this way our bodies "talk to us," through what I call "invisible emotional expression." With invisible emotional expression, we are the interpreters of our own bodies. This is in contrast to "visible emotional expression," which is often interpreted by others.

The Expression of the Emotions in Man and Animals

In addition to our physical selves being taken to indicate enduring traits such as physical strength, our bodies also reveal passing information about our emotions, and what we desire in the moment. As Darwin notes in *The Expression of the Emotions in Man and Animals* there is a great deal of commonality in how we and our non-human relatives signal these emotional states. In *The Expression* Darwin aims to close the perceived gap between humans and animals, to further support his theory of evolution by natural selection (Cain 2009). *The Expression* can be thought of as the complement to *The Descent* (Cain 2009), which was published one year prior.

Darwin begins *The Expression* with three chapters that introduce his three "general principles of expression." In the first chapter he focuses on habits that are not under voluntary control; in the second chapter he presents his Principle of Antithesis or opposites; and in the third chapter Darwin discusses emotional responses that are caused by an "excited" nervous system. He then goes on to apply these three principles to a number of cases of emotional expression in humans and non-human animals in the remainder of the book.

In these discussions, Darwin presents cases of expression of emotional states that he believes we share with non-human animals. Of the heart, Darwin writes the following,

> The heart, as I have said, will be all the more readily affected through habitual association, as it is not under the control of the will. A man when moderately angry, or even when enraged, may command the movements of his body, but he cannot prevent his heart from beating rapidly. His chest will perhaps give a few heaves, and his nostrils just quiver, for the movements of respiration are only in part voluntary. In like manner those muscles of the face which are least obedient to the will, will sometimes alone betray a slight and passing emotion. (Darwin 1872/2009: 77)

In this passage Darwin draws attention to those bodily states that are not under the control of the will. These he emphasizes with his first general principle of expression. Darwin writes that when we try to repress certain movements this is belied by those bodily states outside our control: those "which are least under the separate control of the will, are liable still to act; and their action is often highly expressive" (Darwin 1872/2009: 56). Countless metaphors describe situations in terms of the effect on the body. Something might be described as "crushing," "uplifting," "gut-wrenching," "spine-tingling," "heart-breaking," "toe-curling," eye-popping," and so on.[1] We cannot simply choose not to blush, not to sweat, not to have our heart beat rapidly. This is why, as noted in previous chapters, we sometimes take pains to imitate or conceal these automatic responses.

Darwin may have had special first-person insight into the body and the ways that it responds to the demands of the world because his own body seemed always on the brink of failure. Indeed, Darwin had undefined illnesses that plagued him his whole life (Colp 1977; Barloon and Noyes 1997; Colp 1998; Richards 2017). In an 1865 letter Darwin described his symptoms in the following way,

> For 25 years extreme spasmodic & nightly flatulence: occasional vomiting; on two occasions prolonged for months. . . . .Vomiting preceded by shivering, hysterical crying[,] dying sensations or half-faint, & copious very pallid urine. Now vomiting & every paroxys[m] of flatulence preceded by ringing of ears, rocking, treading on air & vision. Focus & black dots[.] All fatigues, specially reading, brings on these Head symptoms[,] nervousness when E[mma]. Leaves me. (Richards 2017: 52)

It reads very much like Darwin had panic attacks and anxiety, with all the concomitant bodily symptoms (Colp 1977; Barloon and Noyes 1997; Colp 1998).

External bodily states play an important role in how we make hypotheses about the mental life of others. A psychotherapist may, for instance, make the assessment that a patient is in denial if they laugh while describing the death of a parent (Jewett 1982). We can make an assessment of Darwin's mental state based on the symptoms he described in 1865. In *The Expression of the Emotions in Man and Animals* Darwin draws on this connection between the physical manifestations of emotional states (vomiting, shivering, hysterical crying) and the underlying emotion itself (dying sensations, anxiety, PTSD). He has later been diagnosed with panic disorder by MDs examining the historical record for evidence of his symptoms (Colp 1977; Barloon and Noyes 1997; Colp 1998).

Darwin wisely wrote his book not about the emotions themselves—which we cannot know because we cannot look into the minds of others—but on their *expression*, the visible, outward signs that we take as evidence of some emotional state, in ourselves and others, including animals. If we are attuned enough to notice them, we know that in ourselves these correspond to a certain internal experience; it is not a huge leap to assume the same holds for others.

Sometimes our bodily, emotional states rise to a level of conscious awareness such that we are able to express this awareness in words to others (see Rosenthal 2006). Sometimes others are already aware of what we state verbally because they have observed changes in our body. There is a wide range of positions on the relationship between emotions, consciousness, affect, and utterances. We have some words that pick out an internal, subjective state that others cannot directly observe. Wittgenstein writes that if humans did not show outward signs of pain such as groaning or grimacing "it would be impossible to teach a child the use of the word 'tooth-ache'" (Philosophical Investigations § 257). This account makes central the ways we *show* some of our emotional states.

There are a number of distinct philosophical theories about how we ought to treat the relationship between emotions and their expression. Some theories of emotion place the subjective affective phenomenology—not its visible manifestation—at the center. Philosopher Jesse Prinz has argued that emotions are "embodied appraisals" and that all emotions "potentially occur with feelings of bodily changes" (Prinz 2006: 91). Prinz is also explicit to note that in his view "all emotions can be conscious" (Prinz 2006: 91) but does not claim that all emotions *must be* conscious all the time (Prinz 2006: 201–2). Others defend the cognitive view of emotions—that to be in a mental state such as fear is to be consciously experiencing a perceived danger (LeDoux 2017: 303). At some point this becomes a terminological debate, but the fact that there are physical manifestations of emotions—which we may ourselves observe—and thus *realize* we are in a certain emotional state, seems to lean in favor of the Prinz view. On Prinz's view an outside observer can know that some agent is in some emotional state that they are unaware of—just as a mental health professional might be able to say that Darwin had anxiety, on the basis of his description of his symptoms.

Drawing on the Body in Literature

The relation between emotions and visible expression is a topic that has been explored by many authors of fiction and non-fiction. In illustrating this connection in *The Expression* Darwin draws on these literary sources. Indeed, Darwin ends *The Expression* with a quote from Shakespeare's *Hamlet* Act II Scene II as his final word (Darwin 1872/2009: 334). He writes,

> Is it not monstrous that this player here,
> But in a fiction, in a dream of passion,
> Could force his soul so to his own conceit,
> That, from her working, all his visage wann'd;
> Tears in his eyes, distraction in 's aspect,
> A broken voice, and his whole function suiting
> With forms to his conceit? And all for nothing!

Drawing still on Shakespeare Darwin calls attention to the expression of emotions resulting from shyness. He sees the lowering of one's eyes as an expression of shame quoting *Titus Andronicus* Act II, "Ah! Now thou turns't away thy face for shame" (Darwin 1872/2009: 297). Expression of emotion that is remarked upon by the characters in a play is a window into the mental life of the characters. This also helps serve as a sort of instruction to the actors who play these roles for what they should do with their bodies. And when we read the text rather than see it performed on a stage, these sorts of comments allow us to imagine the movements of the characters.

On the topic of blushing Darwin references a passage from *Romeo and Juliet*,

> Thou know'st the mask of night is on my face;
> Else would a maiden blush bepaint my cheek,
> For that which thou hast heard me speak to-night. (Darwin 1872/2009: 308)

Darwin comments on this bit of Juliet's dialog in his text, links blushing to shame, and draws on anecdotes he has been told, writing, "when a blush is excited in solitude, the cause almost always relates to the thoughts of others about us—to acts done in their presence, or suspected by them; or again when we reflect what others would have thought of us had they known the act" (Darwin 1872/2009: 308). In addition to quoting Shakespeare Darwin also refers to other texts, including the Bible. He draws on this range of sources to show that this link between certain bodily states and certain emotions has been utilized in works of fiction and non-fiction across times and cultures.

Expression and Adornment in Darwin's Work

Darwin in *The Expression* discusses the ways our bodies signal emotional states. With his discussion of blushing, drawing on the case of Juliet and other personal anecdotes, he takes this one step further to consider how this expression may differ by sex. He thereby connects the focus of *The Expression* up with the topic of *The Descent*, his book on sexual selection published one year earlier. In this comment on blushing Darwin writes,

> We can understand why the young are much more affected than the old, and women more than men; and why the opposite sexes especially excite each other's blushes. It becomes obvious why personal remarks should be particularly liable to cause blushing, and why the most powerful of all the causes is shyness; for

shyness relates to the presence and opinion of others, and the shy are always more or less self-conscious. (Darwin 1872/2009: 317)

If we take Darwin's point that blushing is more common in the young and in women, we could see why this sort of thing that would be selected for.

In connecting *The Descent* with *The Expression* in this discussion of blushing Darwin brings us closer to one of the main points I have argued for in the book: bodily adornment can be used to create the appearance of natural meaning, which others then draw different inferences from. This is what I have called imitation of natural meaning. We can see why such visible emotional expression would have co-evolved to change the behaviors of those who see and interpret our bodies.

To link this section with the discussion of imitation of natural meaning, it also becomes evident, if Darwin is right, why one would use makeup to imitate these natural features. It is not a stretch to say that the makeup product we call "blush" imitates the natural bodily state of blushing. In a world that prizes youth and femininity women are pushed toward imitation of natural meaning.

The Mind Informs the Body

Let me now turn from discussing visible emotional expression, to discussing invisible emotional expression. Although it remains a mystery how consciousness as we experience it relates to the body or brain as physical stuff,[2] it is clear that, of course, the mind as consciousness does impact the body. This is as obvious as the fact that we have the ability to have a mental state, such as the desire to raise our right arm, and then can we move our right arm. You can do this right now if you wish. But the messages that are sent from ourselves as "thinking things" to our bodies are not always within our conscious control.

Sometimes we are in a mental state that has deleterious effects on our body, and as much as we might wish to simply stop our bodies from reacting in this way, we cannot. Darwin experienced this as well, as evidenced by his self-described bouts of "extreme spasmodic & nightly flatulence," "vomiting," "shivering," and "hysterical crying" (Richards 2017: 52).

Of course, this is not unique to Darwin. Since the time of Darwin the correlation between mental states and bodily states—and in particular between stress and immune response—has been well established. In work on the physiological effects of personal relationships Janice Kiecolt-Glaser and colleagues have found that "men and women who had recently undergone a marital separation or divorce had poorer immune function than demographically matched married individuals" (Kiecolt-Glaser et al. 1998: 657). Additionally, Kiecolt-Glaser and colleagues have found that lonelier medical students had weaker immune responses than those with more social support (Kiecolt-Glaser et al. 1984; Glaser et al. 1992). Helen Fisher and colleagues have done fMRI (functional magnetic resonance imaging) research on a small sample of those who had recently experienced rejection in love and found that when viewing a photo of their former partner they showed activation in the areas of the brain associated with cocaine addiction (Fisher et al. 2010). Fisher and colleagues use this as a way to help understand the obsessiveness associated with rejection by a romantic partner (Fisher et al. 2010).

The Body Informs the Mind

These are important cases to note but the mechanisms underlying the relationship between our minds and our bodies is not unidirectional. It is not simply that our minds feel a certain way and this is revealed through the body. Our bodies also send information to our brain. As noted previously, when participants in a study received undetectable subcutaneous Botox rather than a placebo between their eyes—where the brow furrows in concentration and anger—this resulted in them feeling happier (Finzi 2013). In this way, our bodies convey meaning not just to others but to our own minds. This is never more apparent that when we consider the bodily changes that may alert the conscious mind to emotional distress.

A Psychology of the Body: PTSD

Certain schools of Western psychology and psychotherapy[3] have been increasingly focusing on the embodied nature of certain emotional states (Levine 1997; Van Der Kolk 2015). Bessel Van Der Kolk, a physician and trauma specialist, cites Darwin's *The Expression* in his bestselling book on the relationship between the body and trauma. Van Der Kolk writes, "Darwin goes on to observe that the fundamental purpose of emotions is to initiate movement that will restore the organism to safety and physical equilibrium"

(Van Der Kolk 2015: 75). Darwin believed that the "heart, guts, and brain communicate intimately via the 'pneumogastric' nerve, the critical nerve involved in the expression and management of emotions in both humans and animals" (Darwin 1872/1998: 71–2; Van Der Kolk 2015: 76). Of this passage Van Der Kolk writes,

> The first time I encountered this passage, I reread it with growing excitement. Of course we experience our most devastating emotions as gut-wrenching feelings and heartbreak. As long as we register emotions as primarily in our heads, we can remain pretty much in control, but feeling as if our chest is caving in or we've been punched in the gut is unbearable. We'll do anything to make these awful visceral sensations go away, whether it is clinging desperately to another human being, rendering ourselves insensible with drugs or alcohol, or taking a knife to the skin to replace overwhelming emotions with definable sensations. How many mental health problems, from drug addition to self-injurious behavior, start as attempts to cope with the unbearable physical pain of our emotions? If Darwin was right, the solution requires finding ways to help people alter the inner sensory landscape of their bodies. (76)

It is this goal of treating psychological problems by focusing on the body to which Van Der Kolk has devoted his career.

Some schools of therapeutic treatments for mental health problems, such as those advocated for by Van Der Kolk, focus on the physical, bodily symptoms that accompany conditions such as PTSD. Someone who has experienced trauma has physical symptoms such as "intolerable sensations in the pit of their stomach or tightness in their chest" (Van Der Kolk 2015: 210). These feelings can cause patients to dissociate from their physical selves, a response that can be mitigated by the practice of bodywork (Van Der Kolk 2015: 249).

Van Der Kolk's focus on PTSD as a condition of the body as well as the mind differs from how PTSD has been treated in combat veterans in America. Between 2001 and 2011 the VA (Department of Veterans Affairs) spent approximately $1.5 billion on the antipsychotic medications Risperdal and Seroquel (Van Der Kolk 2015: 228–229). The Department of Defense and VA often provide such medications without other forms of therapy (Van Der Kolk 2015: 228). A 2001 paper showed Risperdal was no more effective than placebo at treating PTSD. It might not be surprising then, that on average, approximately twenty American veterans die by suicide each day, a rate that is 50 percent higher than the general population (Steinhauer 2019). Focusing on the body instead of medication can lead to more effective PTSD treatment (Van Der Kolk 2015).

A Psychology of the Body: Depression

The strong connection between certain mental health conditions and the body seems to fly in the face of understanding such conditions solely in terms of brain chemistry. Ailments such as fibromyalgia, a disorder "characterized by chronic, widespread pain of unknown origin," can be treated with both psychotherapy and medicine (Berenson 2008). Some in the medical profession believe fibromyalgia not to be a "real" disease, but a physical response to depression (Berenson 2008). Such cases confound any attempt at drawing neat distinctions between the mind and the body (Van Der Kolk: 2015: 293; Ming and Coakley 2017). In the treatment of depression, exercise has been shown to be an incredibly effective treatment, on par with antidepressants and psychotherapy (Craft and Perna 2004).

Depression affects just under 10 percent of the U.S. adult population each year; it is also the leading cause of disability in the country (Craft and Perna 2004). A number of studies have demonstrated that exercise reduces depression at rates higher than being assigned to a social support group, or to a control group, and no significant difference was shown between aerobic or non-aerobic exercise (Doyne et al. 1987; Craft and Perna 2004). Research supports the hypothesis that it is in the release of neurotransmitters such as serotonin and dopamine during exercise that depression is curbed (McMurray et al. 1988; Craft and Perna 2004). A "self-efficacy" hypothesis has also been proposed: that the ability to exercise increases a participant's feeling of competence and this helps curb their depression (Bandura 1997; Craft and Perna 2004). These hypotheses are not mutually exclusive.

Perhaps one of the most remarkable and surprising things to come out of recent research on depression is the connection between depression and gut bacteria. A recent study has demonstrated that a number of species of gut bacteria are not found in people with depression (Valles-Colomer et al. 2019; Pennisi 2019). In the 2009 study, 1,054 participants were categorized as depressed—either because they had been diagnosed as such or because they did poorly on a quality of life survey (Valles-Colomer et al. 2019; Pennisi 2019). Approximately 20 percent were classified as depressed or with a low quality of life. In these, two microbes were missing: *Coprococcus* and *Dialister* (Valles-Colomer 2019; Pennisi 2019). As the authors write, "Communication along these lines has been suggested to be bidirectional, with the gut microbiota playing an active role in processes linked to brain development and physiology, psychology and behavior" (Valles-Colomer 2019: 623). Valles-Colomer and her coauthors write, "neural, endocrine and immune communication lines tightly link the human gut microbiota with the host central nervous system" (Valles-Colomer 2019: 623). If we think of the gut as "my" stomach then this would constitute a case of the body having meaning to itself. If we think about this from the perspective of the gut bacteria, we are a "host." If we see ourselves as hosts for a foreign species then such messages are coming from another creature (that happens to live inside our colon) to us.

Again, it is not understood what the mechanism underlying the connection between these bacteria and the mind might be. It might be that certain gut bacteria is correlated with dopamine (Valles-Colomer 2019). One way to change gut bacteria is through a fecal transplant. It certainly is not a glamorous thing to imagine, but if bacteria is missing in one person this is a way to introduce it.

Fecal transplants have been remarkably effective in treating a deadly condition known as C. diff, which is increasing in numbers and this rise parallels the use of antibiotics (Jacobs 2019). Fecal transplants are effective in 80 percent of C. diff cases; "a single dose . . . can bring patients back from the brink of death" (Jacobs 2019). The FDA has not approved such treatments but is not enforcing regulations against them, essentially giving the unofficial "green light" to these incredibly effective fecal transplant procedures (Jacobs 2019). In addition to the depression research, fecal transplant research is being conducted in connection with obesity, autism, ulcerative colitis, Alzheimer's, and Parkinson's disease (Jacobs 2019; Valles-Colomer 2019). If a correlation is found perhaps in the future we will see fecal transplants being used to treat these conditions as well.

Our Body Talks to Us

When considering the relationship between the self and the body the metaphysics quickly gets complicated. We might say that "our body talks to us." That depends what we count as the body and what we could as "us." If we adopt a Cartesian, dualist position that the self is the conscious mind and the body is other then this is coherent. If the self is taken to be the body then perhaps it makes more sense to say that the body does things that sometimes rise to the level of awareness. This can happen through introspection, and emotional awareness is something that must be learned. For instance, in psychotherapy a patient can be trained to recognize and identify with the physical manifestations of emotion that are available to introspection (Levine 1997; Van Der Kolk 2015).

Coming Back to Language: Affect and Utterances

Sometimes our bodily states are revealed not through introspection but through an external observer commenting on our facial expression or utterances. This can be a trained professional such as a therapist, or an untrained layperson. To a certain extent we are all constantly processing these nonverbal cues from others. Sometimes our interlocutors inform us of what they observe in us, in a way that is surprising. We might be informed that we were giving someone a rude look, told to smile, or asked what's wrong.

Consider the following linguistic example of emotional expression, from philosopher of language István Kecskés,

> (D) Roy: Are you okay?
> Mary: I'm fine, Roy.
> Roy: I would have believed you if you hadn't said "Roy." (Kecskes 2014, 2016)

When we read this case, as competent speakers of the language and emotionally tuned in, we know that Mary is not really fine. As Kecskés draws attention to with this example, there is something about stating someone's name at the end of such a sentence that expresses displeasure. A noteworthy thing about this case is that it may or *may not* have been Mary's intention to convey her displeasure here. In fact, Mary's intention is not relevant to the determination that she is not fine. When Roy points out that Mary has just expressed displeasure with this utterance she may be surprised.

There are also cases where some expressed content is stated verbally but requires a clinically trained or more observant hearer to pick up on. The following is taken from philosopher Anne Bezuidenhout (2001),

> A young woman Marie, who is in psychotherapy because she is suffering from *anorexia nervosa*, tells her therapist that her mother has forbidden her to see her boyfriend. Referring to her mother's injunction, Marie utters:
> [1] I won't swallow that
> Here "swallow" is being used metaphorically, and Stern suggests that the content of Marie's utterance (the proposition she expressed) can be paraphrased as

[2] Marie won't accept her mother's injunction.

> Given her eating disorder, it seems significant that Marie chose to frame her comment about her mother's injunction by using the word "swallow." But once we've accessed the metaphorical interpretation it seems that we've lost the echoes of meaning that might connect what she is saying to her eating disorder and hence to any problems that she might be having with her mother connected to this disorder. (Bezuidenhout 2001: 33–34, italics in original)

As Bezuidenhout points out in this passage if we interpret metaphors in terms of their literal content then we miss out on shades of meaning that seem to be conveyed by the specifics of the metaphor used. It seems we ought to say that Marie expressed something about her eating disorder here, although she may not have consciously intended it (Johnson 2019). We see that there is the possibility for unconscious emotional expression in our utterances, as we find with our bodies.

The issues that arise in Darwin's *The Expression* are relevant to the body, yes, but also to linguistic meaning. Throughout most of this book I have been arguing for the ways that theories from philosophy of language can help us understand meaning in the body. In this section we have seen how starting from a notion of bodily expression can help us understand meaning in linguistic utterances (see also Johnson 2019). Sometimes remembering that we are beings with bodies that have bodily, emotional affective states that are interpreted by others can be illuminating.

What Hoodies Mean

When we think about how others read our emotions, as I have been doing in this chapter, it is important to note that this is not the same for people of all races, genders, and bodies more broadly (Zebrowitz et al. 2010). Some bodies are interpreted as being inherently more angry or violent than others. This has tragic roots spanning back to and through the time of Darwin's contemporaries, such as his cousin Francis Galton, a founder of the disgraced theory of eugenics and George Combe, author of *System of Phrenology* and *The Constitution of Man* (Darwin 1871/2004; Desmond and Moore 2009). Phrenologists believed that you could know someone's character just by looking at their skull, and that character traits are genetic and immutable. Such theories amounted to thinly veiled "scientific racism," because similarities were "found" between the skulls of "foreign" people and those of white murderers (Desmond and Moore 2009: 27–38).

Of course, racism in what bodies are taken to mean continues today. In 2012, seventeen-year-old Trayvon Martin was shot by George Zimmerman while walking home from a convenience store with skittles in his pocket (Blow 2012). Prior to murdering him, Zimmerman had called the police and reported that Martin looked "real suspicious" (Blow 2012). This teenage boy had been wearing a hooded sweatshirt.

Hooded sweatshirts, or "hoodies," are comfortable and practical garments. They usually have a large pocket in the front that can warm hands as well as hold a number of items including skittles or cell phones; the hood can be put on for warmth, protection from the rain, and tied tight if necessary for protection from the wind. When the hood is down it provides a comfortable support on the back of the neck.

I have a favorite black hoodie of my own. I bought it at a truck stop after getting deluged while on an open airboat tour in the Everglades. My cousin and I got caught in the downpour and our tour group had stood shivering on a small island in drenched clothes for an hour or so waiting for the rain to let up. Driving out of the Everglades in sopping clothes we stopped to warm up and hopefully buy something dry to wear at a truck stop; it is there I found my black hoodie. Perhaps it is partly because of that first encounter with the garment—the way that it provided physical warmth and something dry when I'd been freezing in a T-shirt in the rain—but this is one of my favorite items of adornment. I usually wear it not for how it looks but for how it makes me feel. When the temperature is a little cool, if I am not in a professional environment, and want to be as comfortable as possible I reach for this hoodie. It always seems to make me feel a little bit better, almost safe.

It is this type of garment that Trayvon Martin was wearing on the night he was shot, and it sparked debate. In a bit of outlandish commentary, *Fox News* pundit Geraldo Rivera suggested that the hoodie was as much to blame for Martin's death as his actual assailant. Rivera said,

> I believe George Zimmerman, the overzealous neighborhood watch captain, should be investigated to the fullest extent of the law and, if he is criminally liable, he should be prosecuted. But I am urging the parents of black and Latino youngsters, particularly, to not let their children go out wearing hoodies! I think the hoodie is as much responsible for Trayvon Martin's death as George Zimmerman was. (Jeffers 2012: 129)

Rivera later apologized for this position, after facing backlash from a number of people, including his own son (Jeffers 2012: 129).

The ways certain bodies are deemed to look "angry" or "real suspicious" is tied to power and policing—from literal policing as was seen with Zimmerman and the Stonewall Riots, to more subtle ways such as a bouncer not letting a man with a flat-brim hat into a club. Bodies are policed because of the gender and racial categories they are placed in, and then, when certain garments become associated with these categories, the garments themselves are policed, thought to be "bad," "dirty," or even "wrong." It is striking to note that the realm of the moral or immoral can extend even to something as seemingly benign as a piece of bodily adornment.

Can We Change This?

In his discussion of Trayvon Martin's hoodie, philosopher Chike Jeffers asks whether or not there was something to Geraldo Rivera's statement that black and Latino parents should stop their children from wearing hoodies. He breaks this question into parts, stating that this ethical question has a prudential element. The prudential aspect has to do with what "might be prudent in avoiding unnecessary harm" (130). He writes, "If we had reason to believe wearing a hoodie made it almost certain that one would be subjected to potentially fatal violence, then the question would be wholly prudential: it would be beyond obvious that it is imprudent to wear a hoodie and no further deliberation would be necessary" (Jeffers 2012: 130). We can see parallels to the "prudential question" were also encountered in the discussion of what women wear and sexual violence as in the SlutWalk debate, a connection that Jeffers makes explicit in the piece (Jeffers 2012: 130).

In his discussion of hoodies in the context of Trayvon Martin, Jeffers writes that

> the question of whether hoodies should be avoided thus emerges: insofar as what hoodies communicate when worn by some black kids is a message of opposition in response to the alienation of being black and economically marginalized in a racist society, we should strive *to undo the conditions under which black kids wear hoodies in this way.* (Jeffers 2012: 135)

What Jeffers himself concludes is that "we should accept the wearing of hoodies as part of black youth culture and even applaud those who express themselves this way while exploding stereotypes through their pursuit of excellence" (Jeffers 2012: 137). Similar to the attempts at change that were seen with SlutWalks, there was a hoodie-wearing march called "Million Hoodies for Trayvon Martin" that took place in New York City in 2012 (Jeffers 2012: 137). In a similar act of protest, in the House of Representatives, US congressman and former Black Panther Bobby Rush took off his suit jacket to reveal a hoodie and was cut off and escorted out for violating the dress code (Jeffers 2012: 137; Walsh 2012). In his speech on the House floor, given over the pounding sound of the gavel, Rush said, "Just because someone wears a hoodie does not make them a hoodlum," making his call for change clear (Walsh 2012).

Meaning of Garments Changes with the Wearer

One thing that this discussion draws attention to is that meaning of bodily adornment can change with the body of the wearer. This brings us to another important point that was left out of Barthes's attempt to create a semiotics of the meaning of garments in *The Fashion System*: *who* is wearing the garment. As I mentioned in Chapter 2, in his analysis Roland Barthes worked with French fashion magazines from the late 1950s. Due to working with this dataset, all the models were thin, young, white women. As we have seen, Barthes himself acknowledged a number of problems with his analysis. This is another problem that arises from limiting his discussion to these bodies.

Consideration of why Barthes may have discussed only these bodies takes us into a bit of his personal history. Barthes was a closeted gay man when he was working on *The Fashion System* in the 1950s (Saint-Amand et al. 1993). For better or for worse, decorative men's fashion has a long history of being associated with homosexuality (McNeil 2015). Barthes may have steered clear of discussion of men's fashion for reasons of justified self-protection in the climate of his day. With Barthes's discussion of solely women's bodies we may see "the workings of repression in his writing" (Saint-Amand et al. 1993).

Barthes's theory ran into enough theoretical problems aside from this, but it is an important further point that must not be overlooked. When we think about the fact that garments are not just garments on their own, but garments on some body, we can see even further the difficulty of treating adornment as a system of fixed meanings. Even if we could say a wool skirt signals professionalism, considering these only on this, young, thin, white, French women, this clearly would have a different meaning if worn by a man (in most parts of the world in 2021). And it is exactly because of the ways we view certain garments as acceptable only on certain bodies that the Stonewall Riots were started. This day is still today marked by the presence of gay pride parades and other events around the world. These facts borne out by consideration of garments like hoodies and skirts bring to the fore a further problem with treating bodily adornment within a semiotic framing.

Adornment and Perception of Race

We have seen through these case studies that the meaning of adornment can change with the wearer. There has been empirical research conducted that is relevant to this topic. A 2011 psychology study considered the inverse link between race and adornment (Freeman et al. 2011). Can a garment change our perception of the race of the wearer? In the study the authors conclude that there is a link between adornment—as it is taken to signify status—and perceptions of race.

In the study, psychologist Jonathan Freeman and his team presented subjects with two continua of faces manipulated on a computer to show measurable gradations between a white face and a black face (Figure 12). Each face was either in a suit—which was taken to indicate that the wearer is of high status—or a janitor's uniform—which was taken to indicate that the wearer is of low status. They found that participants were more likely to classify a face as white when it

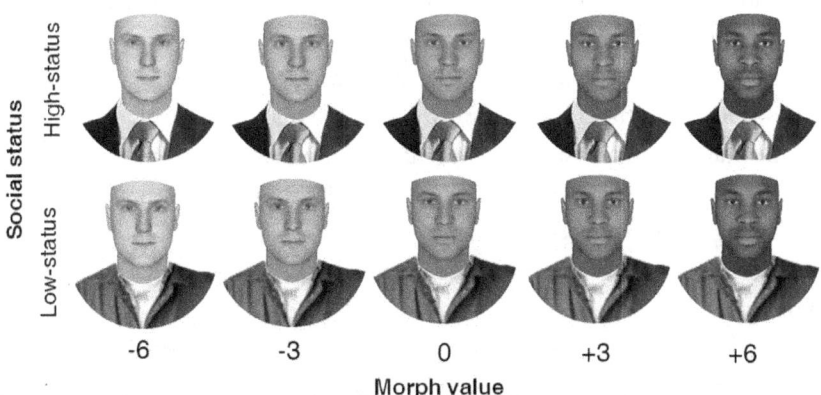

Figure 12 Sample stimuli: 13-point morph continua, with varied race and status attire. We see here the images shown to the study participants in the Freeman et al. article. *Source*: Freeman et al. 2011.

was paired with the "high-status" attire and more likely to classify a face as black when it was paired with "low-status" attire. They take this to be evidence that status cues suggest race.

However, this experimental design takes a number of things for granted that we ought to question. First, we should consider how the two different garments chosen could have affected natural meaning—that is, we should consider whether differences in the visual properties of the two garments make it look as though the faces have different features. The "high-status" attire is a suit consisting of white shirt under a black jacket with a gray tie. In contrast, the "low-status" attire chosen was a royal blue janitor's uniform consisting of a blue button-up shirt over a white T-shirt.

Let us consider the visual properties of these garments in terms of their (1) color and (2) shape. The first point of relevance in consideration of the visual properties of these garments is their color. The "low-status" attire is a very bright color; the "high-status" attire is black and white. It is well established that the color something appears can be greatly affected by the colors next to it. Thus, we ought to wonder what differences in perception of the race of the figures are due to the fact that they appear next to very dark grayscale colors versus next to a bright blue.

The second point of relevance in consideration of the visual properties of these garments is their shape. There is also a great variation in the general shapes created by these garments. As detailed previously, the suit and tie has persevered as a lasting garment at least in part because of the ways it changes how the natural form appears, specifically in that it has an elongating effect on the wearer (Hollander 1994). The janitor's uniform essentially creates a white triangle on its point under the neck, with thinner bright blue triangles on either side of it. The triangle shapes under the face would likely widen its appearance, in stark contrast to the elongation provided by the suit. It could be differences in how the shape and color of these garments affect visual perception that accounts for some of the variation that was found in attributions of race. In this case we should question how the color and the shape of the garments chosen affect the visual perception of the face above and the corresponding associated natural meaning.

Next, we can look at how the non-natural meaning was treated in the case. It is said in the article that the suit is "high-status" attire and the janitor's uniform is "low-status" attire. However, there are a number of questions we should ask about this. First, were subjects able to identify the garment as the type of garment it was meant to be? It is a reasonable assumption that the suit is recognized as a suit. But, what about the janitor's uniform? It seems far from obvious that a blue button-up shirt over a white shirt would be taken to be a janitor's uniform. (Personally, I first thought it was a graduation gown.) Perhaps many of the participants did not recognize the garment as a janitor's uniform.

Then, the next question should be whether or not the garments identified were taken to signify low status or high status as the authors assume. In the age of Steve Jobs and Mark Zuckerberg perhaps the suit no longer signals the highest of status. A 2011 *New York Times* article states that "the bare-bones personal uniform is being seen in some corner offices as the ultimate power suit" (Clifford 2011; see also Friedman 2014; Friedman 2018). And something as specific as how a tie is tied and its fabric can indicate how much it cost. In the illustration used by Jacobs et al. the tie is quite shiny with a large knot with no dimple. And certainly, if viewers did not identify the blue garment as a janitor's uniform there would be no reason to think they thought it was low status. There does not seem to be anything obviously low status about a blue button-down over a white shirt.

The last point of criticism is that the experimenters assumed that the meanings of these garments are fixed across all cases. This is the same assumption that resulted in Barthes's failed attempt to develop a complete semiotics of clothing. That is, the authors thought that they could have two different garments that were with binary independent variables—low-status signifier and high-status signifier—that the dependent variable of the race could be attributed to. However—and this seems even more likely given the results of the experiment—there is reason to think that the way the non-natural meaning of the garment is interpreted varies across contexts (which in this case is the various skin tones of the figure). If, as the authors propose, race is tied to status markers, then it seems a reasonable hypothesis that the meaning of the *clothes themselves* would be interpreted differently in virtue of the perception of the race of the wearer. That is, the perception of how high or low status the garment could be a variable that is *dependent upon* the perceived race of the face. Thus, there could be two changing and dependent variables in the experiment.

In the published experimental results, there is no indication of how these garments were chosen or that any of the points I have raised here were considered. The authors' results are predicated on their precarious assumption that the two garments function as unchanging, binary high- and low-status signifiers. This undefended assumption threatens the significance of their results. I hope this example has shown the value of the framework of clothing interpretation I have presented here; it has allowed us to tease apart a number of questions that were run together with very little attention in a psychology study that made the mistaken assumption that clothing can be understood in terms of fixed codes.

9

On Beauty: Aesthetic Choices, Adornment, and Art

Introduction

In previous chapters I discussed the ways that natural and sexual selection have shaped human and animal bodies, and how we choose to adorn them. I also considered the ways that shame about the body and cultural expectations have shaped norms of dressing. Of course, this is not the whole story—we dress with many aims other than just attracting a partner or staying warm or cool in the elements. We don't always want to be pretty. Sometimes we dress to conceal and to avoid the sexual gaze of others. Sometimes we dress to signal power. Sometimes we dress to show we are punks, or mods, or rockers, or bikers, or what have you.[1] Sometimes we wear a garment with the intent to cause change in the meaning that is attributed to the garments through a process of metalinguistic negotiation. Sometimes we might dress in a garment simply because we are impressed by a soft fabric or a detailed embellishment, because it makes us feel like are wearing a work of artistry. Often aesthetic judgments guide how we choose to adorn. To address this reality—and to return to the matter of the role of beauty in adornment which arose in the disagreement between Wallace and Darwin—in this chapter I will consider how adornment relates to our notions of beauty and art.

Ornithologist Richard Prum—whose account of sexual selection I drew on previously—goes beyond Darwin's claims about beauty and argues that certain animal bodies are art. Prum has a focus on the mechanism of co-evolution, which we have seen previously. He argues that the sexual selection mechanisms that led to things like the peacock's tail are co-evolved aesthetic choices. In this chapter I draw on this aspect of Prum's theory. Although Prum's work is helpful in its characterization of the forces of sexual selection and natural selection, in this chapter I argue that it fails as a theory of art or beauty. Although sexual selection may lead to traits that we judge to be beautiful, this is not sufficient to say that they are art. My argument against Prum does not rule out the possibility that some forms of bodily adornment should be considered art—indeed I will close the chapter by presenting some cases of adornment that I think *are* art— but their being art cannot be merely in virtue of their perceived beauty. In such discussions we must remember that while sexual selection has led to peacock feathers and bird song, sexual selection also leads to other traits that we do not judge to be beautiful, such as the nose of the proboscis monkey. Further, Prum's account fails to exclude the possibility that adorning behaviors are best understood as language-like, which, I will argue, provides support for the account I have presented here. The end of this chapter will include a discussion that brings together a number of threads from elsewhere in the book. In particular I will connect Prum on co-evolution to Grice's work on meaning, and to backlash Darwin faced about his theory superseding the need for a creator to explain the natural world.

Prum's Account

Let me begin the chapter by explaining what Prum has to say about beauty and art. His discussion will be the backdrop for my positive proposal toward the end of the chapter. In a 2013 article in *Biological Philosophy* and in the last few pages of his 2017 book *The Evolution of Beauty* Prum argues that philosophical discourse on aesthetics has overlooked the important role of aesthetic choices in non-human animals. Prum argues for a "biotic aesthetics" that is centered on the co-evolution of aesthetic traits and aesthetic judgments. In what amounts to the "grand finale" of his 2017 book Prum writes that "bird songs, sexual displays, animal-pollinated flowers, fruits, and so on are *art*" (Prum 2017: 336). In a way, the book reads as though it has been building to this point.

Prum's project—of an ornithologist linking research on birds with philosophical problems as he understands them—should be commended. His book is engaging and informative. It is no easy task to engage with literature that is outside our main area of study and thus the principle of charity must be doubly applied in such cases. I do my best to be charitable here, but in this case it does not lead us to the conclusion that Prum is right. Although Prum makes some noteworthy points about co-evolution and the development of what we deem to be beautiful features in nature, overall Prum's theory of co-evolutionary biotic aesthetics does not succeed as a tenable theory of beauty or art.

Animal Art

Prum is far from the first to claim that animals make art. In *The Descent* Darwin has a chapter in which he argues that animals possess many of the same "mental faculties" as *Homo sapiens*, including language, imagination, memories, attention, and art. In particular, in this chapter Darwin writes that canary bird song is art, and draws a parallel with dialects in language. He writes,

> Nestlings which have learnt the song of a distinct species, as with the canary-birds educated in the Tyrol, teach and transmit their new song to their offspring. The slight natural differences of song in the same species inhabiting different districts may be appositely compared, as Barrington remarks, "to provincial dialects"; and with the songs of allied, though distinct species may be compared with the languages of distinct races of man. I have given the foregoing details to shew that an instinctive tendency to acquire an art is not peculiar to man. (Darwin 1871/2004: 108–109)

Note that in this passage Darwin draws attention to the fact that canary bird songs differ by their location, and as he stresses is thus not innate. Darwin implies here that art should be a learned behavior that is not the same for all members of some species.

Perhaps the best candidate for animal art comes from the bowerbird, which Darwin also discusses. Bowerbirds construct structures, called bowers, as a part of their courtship behavior. As Darwin describes them, bowers "decorated with feathers, shells, bones, and leaves, are built on the ground for the sole purpose of courtship, for their nests are formed in trees" (Darwin 1871/2004: 430). In *The Descent* Darwin includes an illustration of a bowerbird with his bower, "These curious structures, formed solely as halls of assemblage, where both sexes amuse themselves and pay their court, must cost the birds much labour" (Darwin 1871/2004: 432). For Darwin, bowers are an illustration of his theory of sexual selection.

Since the time of Darwin, bowerbirds have continued to fascinate researchers (Milam 2011: 89–94). As Jared Diamond describes in his 1982 survey, bowers can be up to eight feet high and are created by eighteen species found in Australia and Papua New Guinea (Diamond 1982). At least six species use fruit or charcoal to color the sticks of the bower themselves and at least four species "orient the bower in a constant compass direction" (Diamond 1982: 99). Between 1965 and 1971 researcher Reta Vellenga banded 940 satin bowerbirds and tracked their behavior for over 15 years (Vellenga 1970, 1980; Diamond 1982). Satin bowerbirds are found in southwest Australia and produce bowers of approximately ten square feet (Diamond 1982: 99). With bowers we see not only evidence of what Darwin called the "charms" of the males but also the "war" for mates. That is, in addition to creating their own bowers, male bowerbirds also destroy the bowers of others. As Jared Diamond writes, "An adult male spends much time repairing his own bower, protecting it from raids by rivals, and attempting to steal ornaments or destroy rivals' bowers" (Diamond 1982: 100). Bowers also meet the condition for art that Darwin implied in the quoted passage earlier, in that the behavior of the male bowerbird is not entirely innate. Further research by Jared Diamond using poker chips to track color preferences has shown that there are different "styles" of bowers found by the same species in different locations (Diamond 1986).

In his book *The Artful Species* Stephen Davies takes a different approach to bowerbirds and argues instead that bowers should *not* be considered art. Davies argues that whale song is the best candidate for animal art. In this argument, Davies cites Jared Diamond on the cultural factors at play in the production of bowers and notes that "young birds must learn local styles by observing the construction of older males" (Davies 2012: 32). He notes that bowerbirds will also sort poker chips and display them among other items (Davies 2012: 33). Indeed, as Diamond saw in his research, bowerbirds seem to have an affinity for a number of brightly colored plastic items. The satin bowerbird, is especially drawn to blue plastic bottle caps (Siossian 2018). In Canberra, Australia's capital, the rings on the top of milk cartons were changed from blue to black to be less attractive to bowerbirds (Siossian 2018). This was necessary because, sadly, these bottle caps present a hazard to the birds, who are prone to getting the rings stuck through their beaks and around the back of their heads. This is fatal if the plastic ring is not removed (Siossian 2018).

Davies writes that the appeal of plastic to bowerbirds "demonstrates that the birds are fussy about color but very indiscriminating in other respects, which shows how inflexible are the biological programs that govern the construction and decoration of their elaborate bowers" (Davies 2012: 33). But why should we assume the satin bowerbirds would reject the plastic, which is, after all a brilliant

On Beauty: Aesthetic Choices, Adornment, and Art 163

Figure 13 A satin bowerbird at Lamington National Park in Queensland, Australia. We see a satin bowerbird here at his bower with a number of his coveted blue plastic items, including a pen in his mouth and bottle caps at his feet. *Source*: Getty Images.

blue? I think this analysis says more about our anti-plastic preference than some eternal truth of aesthetics. It also perhaps taps into the guilt we feel seeing plastic in an animal display, a reminder of human excess polluting the earth. Davies acknowledges that the bowers are shaped by cultural forces and the bowerbird affinity for plastic does not take away from this fact (Figure 13).

Davies's argument in favor of whale song as art is brief and is focused on the fact that such songs are "hierarchically ordered" and constantly change, which is suggestive of "improvisation" (Davies 2012: 33–34). It is not clear if hierarchies and improvisation are necessary or sufficient for animal art in Davies's view. It also is not clear why the bowerbirds' creation process would not be an instance of improvisation.[2]

Prum's Account

The songs of canary birds and whales, as well as the bowers of bowerbirds are animal outputs that are impressive and beautiful to us. Does that mean they are art? Darwin hints at the conditions he sees as what something must have to count as art. Prum's theory is broader than Darwin's brief discussion of art and Prum spends a good deal of time outlining specifically what he sees as art's defining characteristics. This is what Prum calls "biotic aesthetics," which he presents and defends in his 2013 paper and 2017 book. The aim of Prum's proposed biotic aesthetics is to refocus the debate in philosophical aesthetics—which Prum says is currently lacking because of an unprincipled focus on humans. He likens the possible future shift from a "human-centered" aesthetics to his broader biotic aesthetic to the shift from geocentrism to heliocentrism (Prum 2017: 337). Prum argues that aesthetics should involve a broader focus to include the outputs of non-human animals and features of the natural world. These are bold claims indeed. Have philosophers really been this blind to all the forms of beauty found in the natural world around us?

The core of Prum's biotic aesthetics theory is the process of co-evolution. As I've described, co-evolution is a common phenomenon in the natural world, found between two groups that are needed for sexual reproduction, including males and females of the same species, or flowers and pollinators of different species. We've seen how this happens for peacocks and peahens. Similarly, male and female zebra finches, to pick another example Prum discusses, are engaged in a system where the preferences and aesthetic judgments of one sex change and influence the other (Prum 2013). This also can happen across species. For instance, a certain species of hummingbird, the swordbill of the Andes mountains, and a certain species of flower, the passionflower, have co-evolved. The exaggerated beak of the hummingbird, which reaches up to four inches and is longer than its body, is the perfect length to reach into the extended chamber of the flower (Johnson Prum: 2016; see cover image). The chamber of the flower is the length it is because the beak of the hummingbird is the length it is; the beak of the hummingbird is the length it is because the chamber of the flower is the length it is. They co-evolved.

Prum argues that in addition to these animals and plants, human art objects are also a result of this same basic process of co-evolution. That is, in the art world, artists and artistic evaluators are engaged in a system where the preferences and aesthetic outputs each change and influence the other. As I noted in a previous chapter, Prum also takes this to be the case for fashion, where production of some garment and a preference for that garment must grow in tandem.

Prum uses these premises about co-evolution to argue for the following two conclusions: (1) non-human animals make aesthetic judgments of the bodies and behaviors of other non-human animals and plants, and (2) the bodies and behaviors of non-human animals and plants that are the product of such judgments are art (Prum 2013). Prum also has a secondary argument about two different types of beauty in nature that operates independently from these bolder claims. I will present this secondary argument later in the chapter. For now, I will respond critically to the main claim Prum makes in his essay: that biotic aesthetics is a theory of art.

Relevant Questions

As a starting place in considering Prum's position we should begin by distinguishing between the following two questions:

1) Has sexual selection led to certain animal traits that *we* judge to be beautiful?
2) Has sexual selection led to certain animal traits that *animals* judge to be beautiful?

The answer to the first question is undoubtedly "yes." Although it must be noted that, at the same time, sexual selection has also led to many animal traits that we judge to be ugly, repulsive, disgusting, and so on. Think here of the proboscis monkey. The males and females have a "pot belly" that digests leaves in the same way cows do, and the sexually selected-for nose of the males can be up to seven inches long (Koda et al. 2018). Prum's focus on birds and flowers skews the picture of what the overall result of co-evolution is. We must keep in mind that the process that led to the peacock's tail also led to the proboscis. If we are to imagine the feeling of a peahen looking at a peacock as an aesthetic judgment, we should also imagine this as the same feeling a female proboscis monkey would have perceiving a male proboscis. At any rate, it is the same co-evolutionary forces that led to both.

With the second question, "Has sexual selection led to certain animal traits that *animals* judge to be beautiful?," the answer is far from clear. The fact that sexual selection has led to certain animal traits that we judge to be beautiful is not evidence that sexual selection has led to traits that animals judge to be beautiful.

Prum seems to assume an affirmative reply to this second question. He also happens to have Darwin on his side. In Chapter 3 of *The Descent*, Darwin goes through a number of "higher" mental faculties and argues that we share them with animals. One of these is a sense of beauty. He writes that a sense of beauty "has been declared to be peculiar to man" (Darwin 1871/2004: 114). But, he writes, "When we behold a male bird elaborately displaying his graceful plumes or splendid colours before the female, whilst other birds, not thus decorated, make no such display, it is impossible to doubt that she admires the beauty of her male partner" (Darwin 1871/2004: 115). As much as appealing to Darwin gives a sense of authority to any position, on this point Darwin's evidence is rather weak. He overlooks traits that appear ugly to us, such as the nose of the proboscis monkey. Although it does not hold appeal for us as a "display" it is the result of sexual selection (Koda et al. 2018).

Darwin was working at a time with very different scientific standards than those we expect today and was prone to occasionally draw sweeping conclusions on the basis of one amusing anecdote. In this same chapter Darwin also argues that dogs have a sense of humor and enjoy practical jokes (Darwin 1871/2004: 92). In support of this claim, he writes,

> Dogs show what may be fairly called a sense of humour, as distinct from mere play; if a bit of stick or other such object be thrown to one, he will often carry it away for a short distance; and then squatting down with it on the ground close before him, will wait until his master comes quite close to take it away. The dog will then seize it and rush away in triumph, repeating the same manoeuvre, and evidently enjoying the practical joke. (Darwin 1871/2004: 92)

One can't help but think Darwin may have gotten a bit carried away with this. But it doesn't mean that Prum's claim is wrong, just that we need more evidence for it. We do know that he has Darwin, of all people, on his side with this point about animals making judgments about beauty, but this does not settle the matter.

One issue with this debate is that it is not clear what would count as evidence for one answer or another to this question. This is because settling the matter requires making a claim about a non-human animal's internal mental state. We can see what animals *do*, but not what they *think*. As philosopher Thomas Nagel argued in his influential 1974 article "What Is It Like to Be a Bat?" we do not know what it is like to be creatures with other perceptual modalities and states. This certainly complicates things. Modern cognitive ethologists and psychologists do, however, often take an animal's observable behavior as evidence of their mental states, just as we do for pre-verbal infants (Baillargeon et al. 1985; Ristau 1990; Cheney and Seyfarth 2007). So, allowing for this possibility: Can an inference about a judgment of beauty be inferred from behavior? And, crucially, what behavior would be relevant here?

Beauty and Desire

One thing I have learned from discussing Prum's account of beauty both inside and outside the classroom is that most non-philosophers, and undergraduate philosophy majors who have not yet taken aesthetics, associate the term "beauty" not with something transcendental but with female beauty. I have a friend who works in the field of "beauty PR" and this means that she works marketing makeup and skincare products to women—not in creating positive press for the abstract Form of the beautiful. From the tunnel-vision view of working as a philosopher and in aesthetics, it is easy to forget what such terms are ordinarily taken to mean.

Philosophers take beauty to capture something broader than feminine beauty. This approach goes back to Plato, who in *Hippias Major* writes of a discussion between Socrates and Hippias on what is beautiful. Hippias's first attempt is to state "a beautiful maiden is beautiful" (Hippias Major 287e). Socrates later describes this response as "ridiculous" (Hippias Major 288b). As Socrates interrogates Hippias he pushes him not just to state things that are beautiful, but to help in zero in on *the* beautiful, which for Socrates would be a Platonic Form. Socrates points out that not only is a maiden beautiful but so too is a mare. In *Hippias Major* the ideas that beauty is tantamount to "the appropriate," "the useful," "the powerful," "the good," "the pleasing," "the gold-laden" are all rejected. As Socrates concludes at the end of the dialog, "beautiful things are difficult" (Hippias Major 304e).

This idea of beauty as capturing more than just the beauty of a "maiden" continues through more modern aesthetic accounts. For Kant a crucial piece of his account of beauty is disinterestedness—the idea that our judgments of beauty should not include a desire for the thing in question. He makes this clear right at the start of his *Critique of Judgment* writing in the second coda, titled "The Liking That Determines a Judgment of Taste Is Devoid of All Interest" that "if the question is whether something is beautiful, what we want to know is not about the thing's existence, but rather how we judge it in our mere contemplation of it" (Kant 1790/1987: 45). Kant concludes this section by writing, "Everyone has to admit that if a judgment about beauty is mingled with the least interest then it is very partial and not a pure judgment of taste" (Kant 1790/1987: 46). We should in some way be cold, objective, distant from the object we deem to be beautiful. Otherwise, according to Kant we are making a different sort of judgment, one that is no longer a "pure judgment of taste."

Prum's biotic aesthetics—his account of beauty based on sexual selection—is centered on co-evolution, a mechanism that operates in terms of mating choice. As we've seen through canonical texts in aesthetics, an important part of working toward an affirmative answer to the second question delineated earlier—"Has sexual selection led to certain animal traits that *animals* judge to be beautiful?"—would require distinguishing judgments of beauty from judgments of sexual desire.

Turner or Tinder

In the Victorian era when Darwin published *The Descent* such a frank discussion would be far beyond the bounds of propriety. Although it takes us into otherwise taboo subject areas (Prum notes his wife has banned him from discussing duck penises at dinner parties (Prum 2017: 150)), we must not lose sight of the *sexual* nature of sexual selection.

In Prum's 2017 book he has an entire chapter on duck penises (149–181) and an extended discussion (243–254) about why the *Homo sapiens*' penis is so much larger than penises in other primates. Despite his apparent willingness to discuss sexuality, the argument Prum makes maintains more than the vestiges of Victorian propriety.

Prum's argument is based on an analogy between sexual selection in animals and artworld selection in humans—not between animal sexual selection and human sexual selection. Why make this leap and argue that animal sexual selection is akin to a collector at an art gallery? Isn't human sexual selection the parallel to non-human animal sexual selection? Why make a parallel to a patron at an art gallery examining a landscape by Turner rather than to a young woman in South Beach for Spring Break scrolling through her Tinder? It is perhaps less elevated but clearly the next step that must be explored.

Sexual Selection, Sexual Dimorphism, and Our Bodies

If we attempt to look at our own bodies as we look at the bodies of peacocks, peahens, and male and female long-tailed widowbirds, we see that there is less sexual dimorphism between the males and females of our species than in theirs. Although there are obvious differences—sexual organs, average height, and so on (Dutton 2009)—we do not have males with the equivalent of the eighteen-inch tail of the widowbird. And as Darwin observed there are differences around the globe of features that arise through sexual selection. In the female body the waist-to-hip ratio has been presented as one feature that is sexually selected for, as is the fat of breasts—which exceeds what would be necessary for feeding babies (Prum 2017; Richards 2017; Davies 2020). This is something that has direct costs to women because there is research that shows that women with larger breasts are more likely to have breast cancer—the second most common cause of cancer among women in the world (Scutt et al. 1997).

Pornography

The theories of beauty that philosophers such as Plato and Kant put forward do not have the result that a porn star or a model in a cologne ad is an art object. This is because a judgment of beauty, as aestheticians understand it, is not the same as a judgment of sexual attraction.[3] As Kant writes it is *something else*. Trying to imagine the mind of a non-human animal is a difficult endeavor, and we must be ever skeptical of our ability to do so. However, we know that for us there is a difference between a feeling of sexual attraction and the feeling we have when we look at a landscape by Turner. For one thing, there are physiological changes that can accompany one and not the other. These are sometimes measured experimentally. Sexual arousal can be measured by scientific instruments that measure blood flow to the genitals (Trivers 2014); aesthetic pleasure cannot.

It is the feeling we have toward the Turner, that *disinterestedness*, to again draw on Kant, that makes this an aesthetic judgment of beauty. But evolution cannot trade on disinterestedness; evolution has one currency: the genetic transmission that results from a sexual act that leads to offspring. It sounds poetic to say that animals are making judgments of beauty, but what they are actually tracking is something baser: sexiness.

In his work, Prum helps himself to a very broad notion of beauty, accepting pornography into his conception of art. In fact, he includes a range of unorthodox outputs in his theory of art, including toys, games, and advertising in addition to porn. Prum at one point asserts that he does not see why these should not be considered to be art, writing, "Although the classification of these genres as art may be troublesome to some, it is not clear exactly how such genres are actually different from art" (Prum 2013: 825). The problem for Prum's account of beauty is that Kant is right that these are judgments of *something else*. Things like pornography are typically not considered to be art because the emotions, feelings, and desires one has when looking at pornography are quite distinct from the emotions, feelings, and desires one has while looking at a work by Turner. This is a flaw in Prum's account.

Language Not Excluded

Prum does not exclude pornography from his account of art, although I argue he should. It is simply a distinct category. He does attempt to exclude language, although I believe his argument for why language is excluded fails. As we have seen, for Prum art is defined in terms of co-evolution. It is his grounding of art in this process that leads to such an inclusive theory of beauty. But apparently Prum doesn't want it to be too inclusive, and he attempts to exclude language from counting as art.

Prum's attempt to exclude ordinary language from his biotic aesthetics—and thus say that ordinary language is not art—fails. In presenting his argument about language Prum begins,

> Biotic advertisements evolve by a distinct mechanism from other communication signals because they are subject to, and coevolve with, the sensory judgments and evaluations of other organisms. To function, all communication signals must coevolve with their receivers, but not all signals coevolve with subjective sensory/cognitive evaluations. (Prum 2013: 816)

In this passage Prum writes that all communicative signals co-evolve with their receivers. This means that co-evolution alone is not enough to pick out the phenomenon he is interested in. He writes that the special *type* of co-evolution that art has is that it co-evolves "with subjective sensory/cognitive evaluations" (Prum 2013: 816). Prum later writes that his co-evolutionary framework "excludes language in general, in which form and meaning *are* co-evolved but not through evaluation" (Prum 2013: 818).

In order to get to this conclusion, we need an account of what "subjective sensory/cognitive evaluation" is. None of the terms in the description themselves make it clear; to demonstrate why not, let us treat them one by one. By saying that the evaluation is "subjective" Prum could have at least two notions in mind. As Hume presented in his account of beauty (Hume 1757/2008; Carroll forthcoming), "subjective" can have two meanings that are relevant to aesthetics. The first is that the experience occurs in the subject, and is not some feature of the external world. Surely this is satisfied both by language and by art. "Subjective" can also have the meaning that the judgment or experience is unique for each individual. Again, surely this holds in the experience of language. Each person has a number of associations and beliefs about the meaning of some word, such as "Einstein" or "Midtown" (Frege 1892/1949; Kripke 1980; Schiffer 2016), and thus our understanding of language is never the exact same as someone else's. Usually—but not always—there is enough overlap of our understanding of words that sufficient communication is achieved. Language is subjective in both the first and second sense of the term, and so it cannot be the subjectivity that is doing the work here.

What about "sensory/cognitive"? Surely any perception involves both the sensory and the cognitive (this might even be close to the definition of what it means for something to be a perception). And surely language and art both are perceptually based processes. When someone yells "Run!" and I work out what he means and exit the building, I have engaged both my sensory and cognitive capacities.

What about the term "evaluation"? On the face of it, there is nothing in this word itself that provides the necessary wedge. Does he mean something like "receiver's rule" (Godfrey-Smith 2012), "interpretation" (Neale 1992, Carroll 1992), or "perlocutionary act" (Austin 1962)? Unfortunately, Prum does not elaborate on what he has in mind for the "subjective sensory/cognitive evaluation." Ultimately, we are left with the conclusion that Prum has not isolated anything meaningful with his "subjective sensory/cognitive evaluation" condition on co-evolution. He certainly has not yet provided what is needed to set language aside.

A theory that includes bird feathers, pornography, and ordinary language seems not to be a theory of art but perhaps a theory of communication, or cultural transmission. As stated in the discussion of pornography I believe this again is a failing of Prum's theory of biotic aesthetics. Now, this is not to say that art cannot contain language, or that certain forms of language, such as poetry, cannot be art. They certainly can. But the problem for Prum is that his theory doesn't have a non-question-begging way of excluding *any* language. On Prum's account me saying "Hi" to my neighbor and "I'll have a regular coffee with room for milk" at the café count as art. These ought not count as art because they are not special in the way we think art should be (Dissanayake 1995).

Intentions

In an apparent departure from what he has argued up to this point in the paper Prum also considers what the role of intentions will be in his account (Prum

2013). In particular, he responds to the view that art objects require some specific type of intention on the part of their creator, as has been argued by some contemporary philosophers of art (Carroll 1992, 2001). Prum writes,

> Art could be required to be the product of an *aesthetic intention* by an individual or a group of individuals. Aesthetic intention consists of the production, presentation or performance of an aesthetic entity or artifact with conscious regard for its potential for evaluation by another individual. I argue that an aesthetic intention requirement does not clearly exclude biotic phenomena as art. (Prum 2013: 823)

Prum then goes on to give a case of a specific biotic phenomenon, a case of zebra finches that "is *consistent with* the interpretation of aesthetic intention in singing male songbirds" (824; italics mine). The case presented is also consistent with there not being an aesthetic intention. In the zebra finch case even if we *did* see irrefutable evidence of an intention to act we would need further evidence that such an intention was an *aesthetic* intention.

What Prum seems to lose track of in presenting this argument, about, as he puts it, "the aesthetic intention requirement" is that it is just that—a *requirement*, a necessary condition. Prum is right that if some animals have aesthetic intentions then an aesthetic intention requirement would not exclude *all* biotic phenomena. It would, however, restrict his account severely to only those outputs that can plausibly be called the result of such an intention. This would, off the bat, exclude any phenomena in plants and lower-order animals such as mollusks. It would also exclude any species-level patterning or behavior that is a product of genetic inheritance rather than selected-for intentional behaviors of the individual animals. It would, in other words, mean that most of the cases Prum has been presenting as art are not art: bird feathers, the red datura flower, and so on. To say that a view is compatible with something that satisfies a proposed necessary condition is to miss the point.[4]

Prum does not seem to realize the serious consequences of the aesthetic intention requirement. Showing that the aesthetic intention requirement may be met by one creature is not sufficient to show that it can be met by all creatures and outputs of biotic displays. We might see this as a small oversight but Prum uses this zebra finch argument in both his 2013 paper and in the conclusion of his 2017 book. As I stated at the start of this chapter it is commendable for an ornithologist to delve into the world of philosophy, but here we see a fundamental oversight: mistreatment of a proposed necessary condition. This leads to a serious logical flaw in Prum's reply to this objection.

The True Aim of Aesthetics

Throughout his argument Prum never considers aesthetic sensibility or feeling. He writes, "There is nothing special about the sensory systems of humans to support our current privileged position in aesthetics" (814). However, questions about aesthetic judgments are not about the perceptual capacities of our sensory modalities—indeed we fall far behind other animals in each and every one of the capacities of the modalities themselves—but about human minds. This is why aesthetics has historically been focused on humans. Once we see that the core questions of aesthetics are about our human minds, it becomes clear that we humans have cognitive abilities that non-animals lack. For instance, humans have the ability to think about the thoughts of others, often called "mindreading" (Baron-Cohen 1995; Tomasello 2010). If you think, as some philosophers do (Carroll 1992), that mindreading and intention recognition are essential to art, then the absence of mindreading capacities in animals is sufficient to say that animals do not produce art.

Beauty and Co-evolution

Although Prum has not succeeded in his goal to provide an aesthetic theory, his theory of co-evolution provides a helpful framework for conceiving of two different basic ways that beauty in nature develops. As he argues, we can make a meaningful distinction between things in nature that are beautiful because of the judgments of other creatures and those that are not. Of this distinction Prum writes,

> Floral shape, color, and scent are not beautiful in the same way as a sunset or the twinkling stars, because flowers are products of coevolving artworlds consisting of multiple competing plant species communicating to, and coevolving with, the sensory systems of multiple discerning, judgmental individuals of different species of pollinators. By contrast, the color of the sunset and the twinkling of the stars are determined by physical mechanisms that cannot coevolve with our evaluations of them. (Prum 2013: 821)

The co-evolutionary story is one about the way many things in nature came into being. Although it is too much of a leap to say, as Prum wants to, that things like flowers come into being because they were judged to be beautiful and are thus art, they did come into being because of some judgment by creatures that have sensory apparatus that is somewhat similar to ours. By contrast, things like sunsets and stars were not the result of any perceptual apparatus and would have the same properties they have now even if life on Earth never existed.

This story about how these two types of things in nature developed also has implications for the future. Those beautiful things that are the result of co-evolution are constantly changing, and the way they appear now is one of the many ways they will appear in the overall lifespan of the species. Looking past Prum's problematic characterization of this process of co-evolution in terms of "artworlds," this distinction between the two forms of beauty in nature is striking, and could play an important role in philosophical accounts of beauty in nature.

Co-evolution and the Distinction between Natural and Non-Natural Meaning

Prum's work also bears on the previous discussion of natural and non-natural meaning. Recall that one of the examples that Grice uses to explain natural meaning is "clouds mean rain." What relation does Prum's account have to natural meaning? We can consider natural meaning in comparison to animal behavior that reflects co-evolution, such as a connection between red coloring and mating behavior. It is a stretch, of course, to say that an animal makes an inference such as "His red coloring means he'll be a good mate," but this is a way that we can spell out the judgment as *we* make it, on the basis of observing animal behavior.

In Grice's account of natural and non-natural meaning we find two different types of explanation: one that develops with/for an interpreter and another that does not. Clouds do not mean rain because of an interpreter. In contrast, to take an example Grice gives to illustrate non-natural meaning: the three rings of the bell means that the bus is full because of an interpreter. We can also see this division in Prum's account of beauty: there are certain things that evolved to be beautiful because they co-evolved with their receivers, or their interpreters, and there are other things that are just beautiful on their own. Similarly, there are things that just have natural meaning on their own; they didn't evolve to be that way, they don't have any advantage in being interpreted, or because of their receiver's capacity to understand that meaning. A cloud gains nothing by me knowing it is going to rain. A sunset gains nothing by me deeming it beautiful. But a peacock benefits from looking beautiful; a flower benefits from smelling sweet.

Non-natural meaning, like peacock feathers, is a product of a process of co-evolution. As I argued earlier in the chapter Prum's account of co-evolution does not sufficiently exclude language. On the contrary, it allows us to see another way language and adornment are linked: they are both the product of co-evolution. The notion of co-evolution allows us another, and perhaps clearer, way to distinguish between natural and non-natural meaning. Natural meaning did not co-evolve; non-natural meaning did co-evolve.

Co-evolution alone picks out a broader category than non-natural meaning, but if something is not co-evolved it cannot be an instance of non-natural meaning. In his work, Grice was explaining human communication, explaining the behavior of a species—us—that we know can do mindreading and does have the complex mental states of intending an intention. For certain evolved features that are communicative, we know an intention is irrelevant—as in the coloration of a caterpillar.

Now, when we think about meaning in animal cases we cannot determine that we have non-natural meaning. The difficulty in linking Grice with non-human animal vocalizations is that we cannot say conclusively what animal intentions and animal minds are like. All we can observe are behaviors and then we can make inferences about minds based on these. For behaviors such as a baboon cry there is evidence that animals do not possess the mental capacities that would be required to classify them as non-natural meaning (Cheney and Seyfarth 2008). Examples of animal minds don't fall neatly within the categories of Grice's theory; his theory is about us, not non-human animals. Perhaps this could be seen as a limitation of Grice's theory but Grice never set out to explain animal behavior.

We as scientists, or as outside observers, can interpret things in the world and take them to mean things. For instance—to use a case that Darwin and Wallace referred to when they were discussing coloration in birds—blood is red on its own. Blood is not red because its redness means anything. But we, as scientists or doctors, could interpret that blood, based on its redness, to indicate the quantity of some property, say, its hemoglobin levels or some such thing (to make up an example). So, the redness of blood could have meaning on its own, but it didn't evolve to be that way because it was part of any co-evolution with an interpreter. This is the case for many things in the human body that the field of medicine has learned to interpret. Think of pregnancy tests, Covid-19 antibody tests, and so on.

In contrast, a creature like a caterpillar that is brightly colored and poisonous did evolve with an interpreter. Now, does that mean that the body of the caterpillar is an instance of non-natural meaning? No—because in this case there's not any doubt: we know that the caterpillar didn't intend this. The caterpillar's mental state, whatever that looks like, doesn't control its coloring any more than my mental state on any given day controls my eye color.

With the notion of co-evolution we can push to the background the questions about the mental states of animals, but note that they have not totally disappeared. Some modern scientists scoff at Darwin's use of animal judgments of "beauty" to describe what one animal senses when it views an attractive mate (Jones and Ratterman 2009). But note that some mental state, which leads to some creature to choose one mate and not another—is still a part of the theory. Even if we use the term "preference" to describe it, we still are postulating something about animal minds. In both cases we observe some animal behavior, and posit the existence of a mental state on the basis of this.

Art and Bodily Adornment

Let us take a step back, now, from the details of co-evolution and meaning to consider what this discussion means more broadly for the account I have defended in the preceding chapters. One of my takeaways from consideration of Prum's argument is that although co-evolution turns out to be a very helpful concept, we should still not accept the conclusion Prum argues for in his paper and book (Prum 2013, 2017): not everything that co-evolved should be considered art.

It strikes me as quite strange to talk of bird feathers as art except perhaps as a metaphor. It strikes me as even stranger (or perhaps comical) to talk of sexually selected-for human features such as a low hip to waist ratio or a large penis as art. And although I do not believe Prum's concept of co-evolution is helpful with respect to the metaphysical question of what makes something art, it can again be helpfully brought to bear on adornment.

Adornment of course does not have DNA and its propagation is not tantamount to the process of replicating DNA as it is in the natural world. But if we use "evolve" or "co-evolve" loosely to pick out "Darwinian populations" we begin to see that adornment does satisfy Prum's definition of co-evolution (Godfrey-Smith 2011). As we saw earlier, Prum draws a connection between the process of co-evolution in animals and in human fashion, writing that "This evolutionary mechanism is rather like high fashion" (Prum 2017: 40). Prum then notes that "The difference between successful and unsuccessful clothes is determined ... by evanescent ideas about what is subjectively appealing—the style of the season"

(Prum 2017: 40). In the cultural process of certain adornments "looking good" the item evolves as the preference for that item evolves. Take any fashion trend, say leopard print skirts (Kaplan 2019). More and more leopard print skirts are produced. As they do the preference for them grows. The trendy item may even get its own Instagram account (Kaplan 2019). Then this preference leads to even more skirts being produced. Then the preference grows stronger. And, at some point this process has played out and the market becomes oversaturated with leopard print skirts and something new takes its place as the "hot" item.

As I discussed in the chapter wherein I responded to objections to the Gricean view, Ashwell and Langton (2011) have argued that the ways fashion shapes our beliefs is a form of pernicious constraint on wearers. They say we are subject to "aesthetic restrictions" of fashion with respect to what garments and compositions of garments are pleasing (Ashwell and Langton: 2011). Ashwell and Langton argue that those garments that are pleasing we deem to be fashionable, without recognizing the ways in which this judgment was outside of our control—what the authors call a "projective illusion." Ashwell and Langton use this to conclude that we are all in some sense "slaves to fashion." As I argued earlier, this is not unique to adornment. In my previous discussion I said that fashion shared this in common with language. Now, drawing on Prum, I can be more specific: they share in common the process of co-evolution. We cannot step outside this process. The things that we see in the world and our own judgments about what looks good are constantly co-evolving.

Adornment and Art

In this chapter I have argued that Prum is wrong to say that all co-evolved traits are art. This is because "art" picks out something narrower. As I argued in the previous section, bodily adornment is co-evolved. To take the claims I have made here to entail that bodily adornment is never art would be the result of metaphysical confusion.[5] Not all bodily adornment is art, but some is: adornment that meets some higher standard can also be art. My aim here is not to provide a definitive account of what counts as art, but to draw attention to the work of designers whose work may fit some higher standard—whatever that may be.

Think now of a piece of adornment that you would consider to be art. What features does it have? In virtue of what do you consider it to be art? Is it the result of extensive labor and skill on the part of the maker? Is it beautiful? Does it make some sort of statement? Does it have meaning? Is it a part of broader art movements?

Certainly, some fashion designers have considered themselves to be artists. As I noted earlier, Paul Poiret, the Parisian fashion designer who was instrumental in bringing about the end of the corset, stated in 1913, "I am an artist, not a dressmaker" (Svendsen 2006: 91; Koda and Bolton 2007: 150). In his memoirs he wrote, "Am I a fool when I dream of putting art into my dresses, a fool when I say dressmaking is an art? . . . For I have always loved painters, and felt on an equal footing with them. It seems to me that we practice the same craft" (Koda and Bolton 2007: 4). Another fashion designer from a few decades later, Elsa Schiaparelli, was called "that Italian artist who makes clothes" by Coco Chanel.

Schiaparelli's fashions blended in seamlessly with the work of the surrealists at the time (Wood 2007; Baxter-Wright 2012; Martyris 2016; see Figure 14). In 1938 Schiaparelli used long black monkey fur across the front of a sweater, and around the top of a pair of boots, giving them an appearance reminiscent of

Figure 14 A display of Schiaparelli fashions from an exhibition at the Victoria and Albert Museum in London, 1971. We see here Schiaparelli fashions including the tears dress she designed with Salvador Dalí. It is no coincidence that this photo looks almost like one of his paintings. *Source*: Getty Images.

Meret Oppenheim's 1936 work *Le Dejeuner en fourrure,* or *Luncheon in Fur,* where a cup, saucer, and spoon are covered in fur (Martyris 2016). This item causes a visceral response. While writing this and looking at images of the piece, I just had to get up and get a drink of water because my mouth suddenly felt dry, my tongue scratchy. The playful, unexpected nature of the piece is present in Schiaparelli's sweater and boots covered with monkey fur. Many of her other designs also evoke surrealist works (Wood 2007; Baxter-Wright 2012; Martyris 2016). This is no coincidence, because Meret Oppenheim, artist of *Le Dejeuner en fourrure,* worked for Schiaparelli for some time, designing jewelry and accessories (Martyris 2016).

Schiaparelli's designs depicted Freudian erotic fantasies—as in her Lobster and Tears Dresses—as well as nightmares, as seen in a hat she created with bugs crawling over it (Baxter-Wright 2012; Wood 2007). A giant fly brooch the size of a dinner plate seems straight out of a nightmare (Wood 2007; Baxter-Wright 2012). Where a painter might depict such a scene on canvas Schiaparelli put these items on models and eventually on the women who would buy such pieces in her shops.

Schiaparelli's most famous collaborator was Salvador Dalí. With Dalí she designed a dress that looked like bones, a hat that appeared to be a shoe placed on the head, gloves with red nails sewn on top, and many other designs that capitalized on and executed surrealist concepts (Wood 2007; Baxter-Wright 2012). This wasn't simply referencing surrealism in some vague way, but actually doing surrealism; these garments are works of surrealist art.

Surrealism "aimed to change perceptions of the world by exploring dreams, the unconscious mind and the irrational" (Wood 2007). With Schiaparelli's designs the female form became such an object. And the female form itself wasn't some imposition on the work of other surrealists, but a way of animating what was seen already in paintings, such as Dalí's *Three Young Surrealist Women Holding in Their Arms the Skins of an Orchestra,* on canvas (Baxter-Wright 2012: 76). The tradition of fashion as art has carried on through modern designers and perhaps the most convincing case for these as artists might be made for Alexander McQueen and Comme des Garcons (Figure 15).

Additionally, those professionals who do makeup and nails call themselves artists—makeup artists and nail artists. Anyone who has tried to do makeup and nails, or simply watched these practitioners at work knows that an incredible amount of skill—and, if not art at least artistry—goes into professionally doing makeup and nails. They are both very hard to get right. At the highest levels makeup takes hours to put on and can completely transform a face (AuCoin

Figure 15 Rihanna at the Met Gala for Rei Kawakubo/Comme des Garcons: Art of the In-Between, 2017. We see here Rihanna wearing a Comme des Garcons ensemble. Of modern designs, this is the type of bodily adornment that is best qualified to meet the high standard needed for art through adornment. *Source*: Getty Images.

1997). Philosopher Julia Minarik (2021) has argued that we should see some made-up faces—those that are intended to be viewed "*qua* made-up"—as works of art. Nails, too, become more than the "merely" decorative, with technicians developing original designs and sometimes creating individual miniscule paintings on each nail, a sort of piece of mixed media in a set of ten.

Perhaps these feel the most like art when they take us the furthest away from imitation of natural meaning: with the long nails blinged out in jewels, with McQueen heels that take the body far from the "natural" to make it look almost alien—think Lady Gaga's most extreme looks. Perhaps it is in these ways that the body most fits Kant's notion of disinterestedness. It is in these moments that we view the body not as a sexual object to be possessed, but as something that can be made wondrous through artifice—through visible, ostentatious falsification and adornment for adornment's sake.

Beauty through Choice

Let me conclude the book with one final important upshot from Prum's work. Regardless of where one ends up with the question of art, the notion of co-evolution takes us back to debates that have preceded Darwin. The theory of evolution was and still sometimes is perceived as a threat to religion. This is because evolution was taken to undermine teleological arguments for God's existence. This is something Darwin was mindful of (Raby 2001; Prum 2017; Richards 2017).

Teleological arguments are based on the premise that things in the world have a design or purpose. This is then used to argue that God exists because this purpose must be given by God. A famous example of a teleological argument is William Paley's watchmaker argument by analogy (Paley 1863/2009). In the short piece Paley asks the reader to imagine walking along the heath and finding a stone. He writes that a stone has no apparent design and thus we find it unremarkable. Then Paley asks us to imagine finding a watch. A watch, in contrast to a stone, serves a purpose and its parts are there to aid in serving that purpose. Paley writes that if we found a watch we would come to the conclusion that it has a watchmaker, on the basis of its design. He writes that there cannot be "design without a designer . . . order without choice" (Paley 1863/2009: 49). It is left to the reader to understand this argument by analogy and to take the apparent design of things in the natural world as evidence of God's existence.

Darwin's theory of natural selection does in some way provide a solution to how there can be "design without a designer . . . order without choice"; certain features are selected for because they help the creature survive in the natural world. But, as we've seen, Darwin's theory consists not just of a theory of natural selection but of sexual selection as well. What is remarkable, and made more explicit through consideration of Prum's work, is that the theory of sexual selection in some way upholds Paley's maxim. Those features that are the result of sexual selection are the result of a choice. On Darwin's theory of sexual selection it is not God's choice but the choice of the members of the opposite sex who chose a mate based on certain features that were attractive to them. With the theory of sexual selection it is the peahens who are the designers of the feathers of the peacock; it was their choice over millennia that led peacocks to have the features they do. We—all species that have evolved in part because of sexual selection—are the designers teleological arguments called for. We are shaped and we are shapers. In a way we are all that divine force, that artist of the natural world. Each of us—human and non-human animal alike—makes individual choices that are miniscule in the grand scheme of things. But, over time, these miniscule individual choices eventually lead to the shaping of our species, and sometimes—as in the case of the swordbill and the passionflower—other species as well.

Notes

Chapter 1

1   Davies describes here being criticized because of what he is wearing. There was a reason that Davies put on whatever led to him being described as looking "homeless." He made certain choices that led to that day, to that specific adornment. Davies, of course, is not denying that he cares or thinks about adornment, and indeed this quoted passage is at the end of a whole book on it. But there is a relevant history to what he describes here, a reason why he does not wear suits, a reason he would not put a flower in the lapel if he did. I will later comment on the topics of men in suits and men in "ornate" or "fancy" adornment, drawing especially on Hollander (1993), Hollander (1994), and McNeil (2018).

Chapter 2

1   I do think that there could be interesting parallels drawn on the similar ways that *intention* is relevant to meaning in all forms of decoration, and a Gricean treatment of both could be given. See Sperber and Wilson (2015); and Johnson (2019) for a relevant discussion of the difference between showing and meaning. Thanks to Simone Gubler for asking me to elaborate on this point in a presentation. Notably, this presentation was given on zoom, which has caused these categories of the self and the home to blend somewhat in a way they hadn't previously in "regular life."
2   I say "broadly" Gricean because I'm open to the idea that the maxims which govern cooperation for Grice's linguistic theory—quantity, quality, relation, and manner—may be different for dress or call for revision. Although a crude pass at working dress within these categories is quantity—don't wear too few or too many pieces of clothing or accessories; quality—don't lie with your clothes, as in not dressing one's age, or wearing military badges that have not been earned; relation—don't wear things that aren't relevant to the current event, such as wearing your crown to dinner with your spouse; manner—don't wear things that don't suit the occasion, such as jeans to a black tie event, or a blazer to a biker bar.

3 The discussion in this section was first published in *Aesthetics for Birds* (Johnson 2019). Thanks to Thi Nguyen for his helpful editorial comments on that piece.
4 Thank you to an anonymous reviewer for Bloomsbury for raising this point.

Chapter 3

1 Of course speakers are not always cooperating and sometimes lie—the topic of the next chapter.
2 For further details on implicature and the Cooperative Principle, see Johnson (2016).
3 Thanks to Graham Priest for this example. Svendsen (2006: 70) raises a similar example. See Farennikova and Prinz (2011) for more on fashionability in context. For further discussion of what I take to be analogous cases of errors—such as slips of the tongue and malapropisms—in the Gricean tradition, see Unnsteinsson 2017. Again we find parallels with language and adornment.
4 The intimate connection Grice sees between linguistic and nonlinguistic meaning is made clear in a number of Grice's examples. For more on this, see, for example, Grice's example of a man who displays a bandaged leg in response to a squash invitation to mean that he cannot play (Grice 1989: 109).
5 Thanks to Michael Glanzberg for posing a question along these lines at a presentation at the IUC in Dubrovnik.
6 Thanks to the audience at the University of Queensland for their helpful questions at this talk.
7 Thanks to the audience at the Interuniversity Center in Dubrovnik for their helpful questions at this talk.
8 Of course, this is not universal, and some may embrace or even feel morally obligated to flout gendered expectations of dress (Cray 2021).
9 My translation. The Baudelaire text I cite here translates it as "The beautiful is nothing more and nothing less than the promise of happiness" (Baudelaire 1863/1993: 393).

Chapter 4

1 Jonathan Adler (1997) makes much of the distinction between lying and misleading. I do not do so here because I do not think utterers are more morally accountable for what they say than for what they implicate. I believe I am in agreement with Sperber and Wilson (1986) on this point.
2 This section first appeared on *Aesthetics for Birds*.

3   I will discuss Darwin's theorizing about blushing (Darwin 1872/2009: 317) in a later chapter.
4   I certainly do not intend to suggest that such acts are deceptive or entail hiding one's "true self." See my previous comments on artifice. Thanks to Lauren Alpert for helpful discussion on this point, especially Alpert (2014). For very interesting and persuasive discussion of the ethics of cosmetic surgery, see Pitts-Taylor 2007.
5   Thanks to Una Stojnić and Elmar Unnsteinsson for helpful questions about clarifying this point.

Chapter 5

1   There are some certain creatures that *can* change their body and here the question of intentionality is up for consideration. See my discussion of cephalopods in Chapter 8.
2   For instance, in his 1872 *The Expression of the Emotion in Man and Animals* Darwin recalls a story of a shy man who gave an entire "speech" at a dinner party moving his mouth with no sound coming out, where the speaker (allegedly) remained completely unaware he was silent throughout.
3   Prum being the notable exception.

Chapter 6

1   Of course, women are not the only ones who sexually desire men and select them on the basis of preferences; likewise men are not the only ones who sexually desire women and select them on the basis of preferences. Because of the Victorian history of these debates and the focus on genetic transmission between male-female pairings gay desire is often ignored in these discussions. Prum is a notable exception to this and devotes a chapter to this topic in his 2017 book (see Prum 2017: 303–19; see also Farrell 2017; Barron and Hare 2020).
2   For modern scholarship on such questions, see for example Jablonski and Chaplin (2000), Jablonski (2006), Crawford et al. (2017).
3   Note the parallels here with this and the previous discussion of fantailed pigeons.
4   I discuss this point further in Chapter 9.
5   Of course, I am speaking in terms of a very rigid gender binary in terms of standards of dressing, which is assumed in most current office cultures and dress codes. As recently as 2017 while working for the state of Florida as a postdoc at Florida International University I was informed by my colleague Elizabeth Scarbrough that technically we were breaking the dress code by wearing pants. For more on how such norms can and should be challenged, see Cray (2021).

6   There also have throughout history been changing norms about which words are "proper" for women to utter (Smith 2021).
7   For more on a possibly analogous case, "genderfucking," see Cray (2021).
8   Such efforts are, of course, not just restricted to women's bodies. For other analogous marches related to other bodies consider the Million Hoodie March and the Chicago Disability Pride Parade. In a later chapter I will discuss the Million Hoodie March (Jeffers 2012). See Barnes (2016) for a lucid argument that builds a philosophical scaffolding in support of events like the Chicago Disability Pride Parade (Barnes 2016: 185). Barnes writes, "Disability pride says that we may have *minority* bodies, but we don't have—refuse to have—tragic bodies" (Barnes 2016: 186).

Chapter 7

1   This radiating pleat effect is found on some modern ruched "body-con" dresses, which emphasize the fullness of a woman's hips.
2   They write of the ochre that "the phenomenon of ochre is to be interpreted as a persistent tradition handed down through the generations of the use of the color red as an *index* for objects, ideas or events" (507; emphasis added). In discussing the burials themselves they write that "a growing number of researchers now subscribe to the view that intentional burial existed in the Middle Paleolithic . . . if so such burials are, *because of their intentionality, symbolic*" (508; emphasis added). And in respect not to the ochre but the bones they write, "the bones of dead kin are *at least iconic* of the living person in that they point to their referent by physical *resemblance*" (508; emphasis added). We might ask: Does it really make sense to say that there is a physical resemblance between some bones and the person they came from? Do the bones of my grandmother resemble her more than they resemble your grandmother? And on top of this, burials can be intentional without being symbolic. It depends on the type of intention; some intentional acts are intentionally communicative and others are not (Johnson 2017). The color red could very well be an index for "objects, ideas, or events" but more argument in support of this is needed.
3   To consider the difficulties we might consider representations of cats. Does a two-dimensional drawing of a cat really resemble a cat? Do the emoticons ^ⓒ*ⓒ^ or (^._.^)ᴦ resemble cats? Why? Certainly they have very little in common with any cat I've ever met. At the same time, there are features of this emoticon that bear a non-arbitrary relation to the cat, that is, the pointy parts to cat ears, which are also pointy.

Chapter 8

1. Bodily, emotional states may play a larger role in communication than is ordinarily acknowledged. In *Metaphors We Live by* George Lakoff and Mark Johnson discuss the way metaphor "is pervasive, not merely in our language but in our conceptual system" (Lakoff and Johnson 1980/2003: 211). For them, metaphors are not "mere language" but rather Lakoff and Johnson advocate for the view that we have metaphorical concepts. Many of these metaphorical concepts are physical and bodily.
2. The idea that our minds and bodies exist as separable entities can be most clearly found in philosopher René Descartes's dualism. In his famous work *The Meditations* Descartes begins by doubting all knowledge that is based on the senses, which he deems to be unreliable. This has the result that early on in the work the only thing Descartes can be sure he knows is that he is a thinking thing, or consciousness (Descartes 1640/1984). As *The Meditations* continue Descartes regains confidence in his ability to trust the senses. He comes to believe that he is not just consciousness but that he has a body with which he perceives the world. As was pointed out to Descartes by his contemporary Princess Elisabeth of Bohemia, Descartes never takes the necessary steps to connect the mind as a thinking thing back up with the body. If the mind is fundamentally a different thing (ethereal consciousness that can live on after death), she asks Descartes: Then how can it move the body, which is made of matter (Elisabeth 1643/1994)? As she writes, how can something without matter, an "immaterial thing" move the body (Elisabeth 1643/1994: 11–12)? Descartes never gives her a satisfactory answer and Princess Elisabeth concludes that there are "unknown properties in the soul" (Elisabeth 1643/1994: 21). This connection between the mind as consciousness and the body as physical stuff has never been sufficiently resolved and is today known as the "mind-body problem."
3. My discussion here is Western-focused; there have been long traditions outside the West of viewing the mind and body as inextricably linked, such as in yoga, tai chi etc.

Chapter 9

1. For more on this point in particular, see Johnson (forthcoming).
2. We also might want to question the very assumption Darwin made in his characterization of canary bird song. Is culture a necessary condition for art? Couldn't art-making be genetically programmed? Why is culture required for or suggestive of art?
3. See Eaton (2018) for more on the distinction between art and pornography.

4 If it is not yet clear exactly what is wrong with Prum's reply to the aesthetic intention requirement let me provide an analogous hypothetical discussion about the classification of hummingbirds. Let's imagine that a certain misguided philosopher posits the following conditions for something to be a hummingbird:

> Misguided philosopher: A hummingbird is anything that is less than three ounces. This means that my pen, my coaster, and this piece of paper on my desk are all hummingbirds. I do not understand why ornithologists have taken such a restrictive view. [This is analogous to Prum's biotic aesthetics.]
> Ornithologist: Surely for something to count as a hummingbird it must also be a bird. [This is analogous to the aesthetic intention requirement.]
> Misguided philosopher: Yes, it is compatible with my position that birds that are less than three ounces are also hummingbirds. For instance here is a bird that is less than three ounces. It is called a ruby-throated hummingbird. My view does not exclude them. [This is analogous to the discussion of the zebra finch case.]
> Ornithologist: Sigh.

I hope this dialogue makes the contours of the dialectic clear.

5 This is a result we want to avoid. See Dadlez (2021) and Minarik (2021) for discussion of tattoos and makeup as art.

References

Adler, J. (1997). "Lying, Deceiving, or Falsely Implicating." *Journal of Philosophy*, 94 (9): 435–52.

Alpert, L. (2014). "Cosmetics, Self-Portraiture, and the Authentic Self." *SWIPshop*. New York: CUNY Graduate Center. Lecture.

Amat, J. A., Rendón, M. A., Garrido, A., Rendón-Martos, M., Péret-Gálvez, A. (2011). "Greater Flamingos *Phoenicopterus Roseus* Use Uropygial Secretions as Make-Up." *Behavioral Ecology and Sociobiology*, 65 (4): 665–73.

Anderson, L. and E. Lepore (2013). "Slurring Words." *Nous*, 47 (1): 25–48.

Andersson, M. (1982). "Female Choice Selects for Extreme Tail Length in a Widowbird." *Nature*, 299: 818–20.

Ashwell, L. and R. Langton (2011). "Slaves to Fashion." In J. Wolfendale and J. Kennett (eds.), *Fashion—Philosophy for Everyone: Thinking With Style*, 135–50. Blackwell.

Atherton, M. (1994). *Women Philosophers of the Early Modern Period*. Indianapolis: Hackett.

Associated Press. (2018). "Grabbing a Beard? The NHL Has It Covered." *Chicago Tribune*, January 5. Accessed online.

AuCoin, K. (1997). *Making Faces*. New York: Little, Brown.

Austin, J. L. (1962). *How to Do Things with Words*. Cambridge: Harvard University Press.

Bach, K. (1987). "On Communicative Intentions: A Reply to Recanatti." *Mind & Language*, 2 (2): 141–54. Oxford: Blackwell.

Bahn, P. (2012). *Cave Art: A Guide to the Decorated Ice Age Caves of Europe*. London: Frances Lincoln.

Baillargeon, R. Spelke, E. S., Wasserman, S. (1985). "Object Permanence in Five-Month-Old Infants." *Cognition*, 20 (3): 191–208.

Baldwin, J. (1960). "On Being Black in America." *Encounter*. CBC. Television interview with Nathan Cohen. Accessed online.

Balzac, H. (2001). *Lost Illusions*. New York: Modern Library.

Bandura, A. (1997). *Self-Efficacy: The Exercise of Control*. New York: W. H. Freeman & Company.

Banner, L. (1984). *American Beauty*. Chicago: University of Chicago Press.

Barloon, T. J. and R. Noyes (1997). "Charles Darwin and Panic Disorder." *JAMA*, 227 (2): 138–41.

Barnes, E. (2016). *The Minority Body: A Theory of Disability*. Oxford University Press.

Baron-Cohen, S. (1995). *Mindblindness*. Cambridge: MIT Press.

Barron, A. B. and B. Hare (2020). "Prosociality and a Sociosexual Hypothesis for the Evolution of Same-Sex Attraction in Humans." *Frontiers in Psychology*, 10: 2955.

Bartels, M. (2017). "These Birds Like to Wear Makeup." *Audobon News*, July 21. Accessed online.
Barthes, R. (1978). *Image-Music-Text*, trans. French by S. Heath. New York: Hill and Wang.
Barthes, R. (1990). *The Fashion System*, trans. French by M. Ward and R. Howard (eds.). Oakland: University of California Press.
Barthes, R. (2005). *The Language of Fashion*, trans. French by A. Stafford. A. Stafford and M. Carter (eds.). London: Bloomsbury Academic.
Bartky, S. (1998). "Foucault, Femininity, and the Modernization of Patriarchal Power." In R. Weitz (ed.), *The Social Construction of Women's Bodies. The Politics of Women's Bodies: Sexuality, Appearance, and Behavior*, 25–45. Oxford: Oxford University Press.
Baudelaire, C. (1863/1993). "The Painter of Modern Life." In P. E. Charvet (ed.), *Selected Writings on Art and Literature*, 358–89. London: Penguin Classics.
Bauer, A. (2013). "Objects and Their Glassy Essence: Semiotics of Self in the Early Bronze Age Black Sea." *Signs and Society*, 1 (1), 1–31.
Baxter-Wright, E. (2012). *The Little Book of Schiaparelli*. London: Carlton Books.
Belluck, P. (2015). "Chilly at Work? Office Formula Was Devised for Men." *New York Times*, August 3. Accessed online.
Berenson, A. (2008). "Drug Approved. Is Disease Real? ." *New York Times*, January 14. Accessed online.
Bezuidenhout, A. (2001). "Metaphor and What Is Said: A Defense of a Direct Expression View of Metaphor." *Midwest Studies in Philosophy*, 25: 156–86.
Blackmore, S. (2000). *The Meme Machine*. Oxford: Oxford University Press.
Blow, C. (2012). "The Curious Case of Trayvon Martin." *New York Times*, March 16. Accessed online.
Burgess, A. and D. Plunkett (2013). "Conceptual Ethics I & II." *Philosophy Compass*, 8: 1091–110.
Cain, J. (2009). "Introduction." In *The Expression of the Emotions in Man and Animals*, xi–xxxiv. London: Penguin.
Camp, E. (2013). "Why Metaphors Make Good Insults." Philosophy Department Colloquium. The Graduate Center, CUNY. December 4, 2013. Lecture.
Cantalamessa, E. (2020). "Appropriation Art, Fair Use, and Metalinguistic Negotiation." *British Journal of Aesthetics*, 60 (2): 115–29.
Carroll, N. (1992). "Art, Intention, and Conversation." In G. Iseminger (ed.), *Intention and Interpretation*, 97–131. Philadelphia: Temple University Press.
Carroll, N. (2001). "Identifying Art." In Nïel Carroll (ed.), *Beyond Aesthetics: Philosophical Essays*, 75–100. Cambridge: Cambridge University Press.
Carroll, N. (Forthcoming). Classics in the Philosophy of Art. Oxford: Oxford University Press.
Carter, D. (2010). *Stonewall: The Riots that Sparked the Gay Revolution*. New York: St. Martin's Griffin.

Carter, H. (1922/1977). *The Discovery of the Tomb of Tutankhamen*. Mineola: Dover Publications.

Chalmers, D. and A. Clark (1998). "The Extended Mind." *Analysis*, 58 (1): 7–19.

Chandler, D. (2007). *Semiotics: The Basics*. Second Edition. Oxfordshire: Routledge.

Chapman, R. and A. Wylie (2016). *Evidential Reasoning in Archaeology*. London: Bloomsbury.

Cheney, D. and R. Seyfarth (1990). *How Monkeys See the World: Inside the Mind of Another Species*. Chicago: University of Chicago Press.

Cheney, D. and R. Seyfarth (2008). *Baboon Metaphysics: The Evolution of a Social Mind*. Chicago: University of Chicago Press.

Choi, Y. S., Gray, H. M., Ambady, N. (2004). "The Glimpsed World: Unintended Communication and Unintended Perception." In R. Hassin, J. Uleman and J. Bargh (eds.), *The New Unconscious*, 309–33. Oxford: Oxford University Press.

Clark, G. (1969). *World Prehistory: A New Outline*. Cambridge: Cambridge University Press.

Clifford, S. (2011). "Power of Apparel: A Look that Conveys a Message." *New York Times*, March 4.

Colp, R. (1977). *To Be an Invalid: The Illness of Charles Darwin*. Chicago: University of Chicago Press.

Colp, R. (1998). "'To Be an Invalid', Redux." *Journal of the History of Biology*, 31 (2): 211–40.

Conkey, M. (2001). "Structural and Semiotic Approaches." In D. Whitley (ed.), *Handbook of Rock Art Research*, 273–310. Walnut Creek: Altamira Press.

Craft, L. and Perna, F. (2004). "The Benefits of Exercise for the Clinically Depressed." *Prim Care Companion J Clin Psychiatry*, 6 (3): 104–11.

Crawford, N. G., Kelly, D. E., Hansen, M. E. B., Beltrame, M. H., Fan, S., Bowman, S. L., Jewett, E. et al. (2017). "Loci Associated with Skin Pigmentation Identified in African Populations." *Science*, 358 (6365).

Cray, L. (2021). "Some Considerations Regarding Adornment, the Gender 'Binary', and Gender Expression." *Journal of Aesthetics and Art Criticism*, 79 (4): 488–92.

Crossland, Z. (2014). *Ancestral Encounters in Highland Madagascar: Material Signs and Traces of the Dead*. Cambridge: Cambridge University Press.

Curtis, G. (2007). *The Cave Painters: Probing the Mysteries of the World's First Artists*. New York: Anchor.

Dadlez, E. (2021). "Tattoos Can Sometimes Be Art: A Modest Embellishment of Stephen Davies' *Adornment*." *Journal of Aesthetics and Art Criticism*, 79 (4): 499–503.

Danziger, P. (2018). "Why Pearls Are the Perfect Luxury Gem for Millennials." *Financial Times*, December 5. Accessed online.

Darwin, C. (1859/2003). *The Origin of Species*. Kolkata: Signet Classics.

Darwin, C. (1871/2004). *The Descent of Man and Principles of Sexual Selection*. London: Penguin.

Darwin, C. (1872/1998). *The Expression of Emotions in Man and Animals*. Oxford: Oxford University Press.

Darwin, C. (1872/2009). *The Expression of Emotions in Man and Animals*. London: Penguin.

Davé, A. (2018). "Do Beards Absorb Punches? Boxing and MMA Rules." *Twisted Moustache*. Accessed online.

Davies, S. (2012). *The Artful Species: Aesthetics, Art, and Evolution*. Oxford: Oxford University Press.

Davies, S. (2020). *Adornment: What Self-Decoration Tell Us About Who We Are*. London: Bloomsbury.

Dawkins, R. (1976). *The Selfish Gene*. Oxford: Oxford University Press.

Dawkins, R. (1996). *The Blind Watchmaker: Why the Evidence of Evolution Reveals a Universe without Design*. New York: W. W. Norton & Company.

Death Becomes Her: A Century of Mourning Attire (October 21 2014–February 1, 2015). *Exhibition. Metropolitan Museum of Art. Anna Wintour Costume Center*. New York.

Delhey, K. Peters, A., Kempenaers, B. (2007). "Cosmetic Coloration in Birds: Occurrence, Function, and Evolution." *The American Naturalist*, 169 (S1 Supplement 1): S145–58.

Denkel, A. (1998). *The Natural Background of Meaning*. New York: Springer.

Dennett, D. (1995). *Darwin's Dangerous Idea: Evolution and the Meanings of Life*. New York: Simon & Schuster.

d'Errico, F. (2003). "The Invisible Frontier: A Multiple Species Model for the Origin of Behavioral Modernity." *Evolutionary Anthropology*, 12: 188–202.

d'Errico, F., Henshilwood, C., Lawson, G., Vanhaeren, M., Tillier, A., Soressi, M., Bresson, F., Marrielle, B., Nowell, A., Lakarra, J., Backwell, L., Volien, M. (2003). "Archaeological Evidence for the Emergence of Language, Symbolism, and Music—An Alternative Multidisciplinary Perspective." *Journal of World Prehistory*, 17 (1): 1–70.

d'Errico, F. Vanhaeren, M., Barton, N., Bouzouggar, A., Mienis, H., Richter, D., Hublin, J., McPherron, S. P., Lozouet, P. (2009). "Additional Evidence on the Use of Personal Ornaments in the Middle Paleolithic of North Africa." PNAS, 106 (38): 16051–6.

Descartes, D. (1640/1984). "Meditations on First Philosophy." In Cottingham et al. (trans.), The Philosophical Writings of Descartes: Volume II. Cambridge: Cambridge University Press, 1–50.

Desmond, Adrian and James Moore. (2009). *Darwin's Sacred Cause: Race, Slavery, and the Quest for Human Origins*. London: Allen Lane.

DeWall, C. N., Macdonald, G., Webstar, G. D., Masten, C. L., Bavmeister, R. F., Powell, C., Combs, D., Schurtz, D. R., Stillman, T. F., Tice, D., Eisenberger, N. I. (2010). "Acetominophen Reduces Social Pain: Behavioral and Neural Evidence." *Psychological Science*, 21 (7): 931–7.

Diamond, J. (1982). "Evolution of the Bowerbirds' Bowers: Animal Origins of the Aesthetic Sense." *Nature*, 297 (May): 99–102.

Diamond, J. (1986). "Animal Art: Variation in Bower Decorating Style Among Male Bowerbirds *Amblyornis inornatus*." *PNAS*, 83 (9): 3042–6.

Diener, E. Wolsic, B., Fujita, F. (1995). "Physical Attractiveness and Subjective Well-Being." *Journal of Personality and Social Psychology*, 69 (1): 120–9.

Dissanayake, E. (1995). *Homo Aestheticus: Where Art Comes From and Why*. Seattle: University of Washington Press.

Donnellan, K. (1966). "Reference and Definite Descriptions." *The Philosophical Review*, 75 (3): 281–304.

Donnellan, K. (1968). "Putting Humpty Dumpty Together Again." *The Philosophical Review*, 77 (2): 203–15.

Doyne, E. J., Ossip-Klein, D. J., Bowman, E. D., Osburn, K. M., McDougall-Wilson, B., Neimeyer, R. A. (1987). "Running Versus Weight Lifting in the Treatment of Depression." *Journal of Consulting and Clinical Psychology*, 55 (5): 748–54.

Dretske, F. (1981). *Knowledge and the Flow of Information*. Cambridge: MIT Press.

Dutton, D. (2009). *The Art Instinct: Beauty, Pleasure and Human Evolution*. London: Bloomsbury.

Eaton, A. W. (2018). "'A Lady on the Street but a Freak in the Bed': On the Distinction Between Erotic Art and Pornography." *British Journal of Aesthetics*, 58 (4): 469–88.

Eisenberger, N. I., Lieberman, M. D., Williams, K. D. (2003). "Does Rejection Hurt? An fMRI Study of Social Exclusion." *Science*, 302 (5643): 290–2.

Elisabeth. (1643/1994). "Princess Elisabeth of Bohemia: Selections from Her Correspondence with Descartes." In M. Atherton (ed.), *Women Philosophers of the Early Modern Period*, 9–21. Indianapolis: Hackett.

Farennikova, A. and J. Prinz (2011). "What Makes Something Fashionable?" In J. Wolfendale and J. Kennett (eds.), *Fashion—Philosophy for Everyone: Thinking With Style*, 15–30. Hoboken: Blackwell.

Farrel, J. (2017). "How Sexual Selection Drove the Emergence of Homosexuality." *Forbes*, May 7. Accessed online.

Fernandez, M. (2014). "When the Texas Governor Drops the Official State Footwear." *New York Times*, July 2. Accessed online.

Fernandez-Llario, P. (2005). "The Sexual Function of Wallowing in Male Wild Boar (*Sus scrofa*)." *Journal of Ethology*, 23 (1): 9–14.

Ferrell v. Dallas School District. (1968). 392 F. 2d 697, U.S. Court of Appeals Fifth Circuit.

Finzi, E. (2013). *The Face of Emotion: How Botox Affects Our Moods and Relationships*. New York: St. Martin's Press.

Fisher, H E., Brown, L. L., Aron, A., Strong, G., Mashek, D. (2010). "Reward, Addiction, and Emotion Regulation Systems Associated with Rejection in Love." *Journal of Neurophysiology*, 104 (1): 51–60.

Fisher, L. A. (2019). "Rihanna is a Pink Feathered Dream at Crop Over 2019." *Harper's Bazaar*, August 5. Accessed online.

Fisher, R. A. (1918). "The Correlation Between Relatives on the Supposition of Mendellian Inheritance." *Transactions of the Royal Society of Edinburgh*, 52: 399–433.

Freeman, J. B., Denner, A. M., Saperstein, A., Schevtz, M., Ambady, N. (2011). "Looking the Part: Social Status Cues Shape Race Perception." *PLOSOne*, 6 (9): 1–10.

Frege, G. (1892/1949). "On Sense and Nominatum." (Translation by Feigl, H). In A. P. Martinich (ed.), *The Philosophy of Language*, 217–29. Oxford: Oxford University Press.

Friedman, J. B. (2018). "Eyebrows, Hairlines, and 'Hairs Less in Sight': Female Depilation in Late Medieval Europe." In R. Netherton and G. Owen-Crocker (eds.), *Medieval Clothing and Textiles: Volume 14*, 81–112. Suffolk: Boydell Press.

Friedman, V. (2014). "This Emperor Needs New Clothes: For Tim Cook of Apple, the Fashion of No Fashion." *New York Times*, October 15. Accessed online.

Friedman, V. (2018). "It's 2018: You Can Run for Office and Not Wear a Pantsuit." *New York Times*, June 21. Accessed online.

Friedman, V. (2018). "Mark Zuckerberg's I'm Sorry Suit." *New York Times*, April 10. Accessed online.

Gilligan, I. (2019). *Climate, Clothing, and Agriculture in Prehistory: Linking Evidence, Causes, and Effects*. Cambridge: Cambridge University Press.

Gilman, M. (1939). "Baudelaire and Stendhal." *PMLA*, 54 (1): 288–96.

Glass, I. (2020). "691: Gardens of Branching Paths." This American Life. Chicago Pubic Media, January 10. Accessed online.

Glaser, R. Kiecolt-Glaser, J. K., Bonneav, R. H., Matarkey, W., Kennedy, S., Hughes, S. (1992). "Stress-Induced Modulation of the Immune Response to Recombinant Hepatitis B Vaccine." *Psychosom. Med*, 54: 22–9.

Godfrey-Smith, P. (2011). *Darwinian Populations and Natural Selection*. Oxford: Oxford University Press.

Godfrey-Smith, P. (2012). "Review of Brian Skyrms' *Signals*." *Mind*, 120: 1288–97.

Godfrey-Smith, P. (2014). "Signs and Symbolic Behavior." *Biological Theory*, 9: 78–88.

Godfrey-Smith, P. (2016). *Other Minds: The Octopus, the Sea, and the Deep Origins of Consciousness*. New York: FSG.

Goodall, J. (2010). *In the Shadow of Man*. Boston: Mariner Books.

Grice, H. P. (1957). "Meaning." *Philosophical Review*. 66 (3) pp. 377–88.

Grice, H. P. (1989). *Studies in the Way of Words*. Cambridge: Harvard University Press.

Hajjaji-Jarrah, S. (2003). "Women's Modesty in Qur'anic Commentaries: The Founding Discourse." *The Muslim Veil in North America*, 181–213. London: Women's Press.

Hanson, K. (1998). "Fashion and Philosophy." In M. Kelly (ed.), *Encyclopedia of Aesthetics*, 158–61. Oxford: Oxford University Press.

Harrano, L. (2020). "These Real Brides Prove You Don't Need to Wear a Dress to Look Stylish on Your Wedding Day." *Popsugar*, February 19. Accessed online.

Hinton, H. E. (1973). "Natural Deception." In R. L. Gregory and E. H. Gombrich (eds.), *Illusion in Art and Nature*, 96–159. New York: Scribner.

Hiscock, P. (2014). "Learning in Lithic Landscapes: A Reconsideration of the Hominid 'Toolmaking' Niche." *Biological Theory*, 9 (1), 27–41.

Hollander, A. (1993). *Seeing Through Clothes*. Oakland: University of California Press.

Hollander, A. (1994). *Sex and Suits*. Tokyo: Kodansha Globe.

Hoodfar, H. (2003). "More than Clothing: Veiling as an Adaptive Strategy." *The Muslim Veil in North America*, 3–40. London: Women's Press.

Hooks, K. B., Konsman, J. P., O'Malley, M. A., (2018). "Microbiota-Gut-Brain Research: A Critical Analysis." *Behavioral and Brain Sciences*, 42 (E60). Accessed online.

Hovers, E., Ilani, S., Bar Yosef, O., Vandermeersch, B. (2003). "An Early Case of Color Symbolism: Ochre Use by Modern Humans in Qafzeh Cave." *Current Anthropology*, 44 (4), 491–522.

Hughes, K. (2018). *Victorians Undone*. New York: 4th Estate Harper Collins.

Hume, D. (1757/2008). "Of the Standard of Taste." *David Hume: Selected Essays*, 133–53. Oxford University Press.

Jablonski, N. G. (2006). *Skin: A Natural History*. Oakland: University of California Press.

Jablonski, N. G. and G. Chaplin (2000). "The Evolution of Human Skin Coloration." *Journal of Human Evolution*, 39: 57–106.

Jacobs, A. (2019). "Drug Companies and Doctors Battle Over the Future of Fecal Transplants." *New York Times*, March 3. Accessed online.

Jeffers, C. (2012). "Should Black Kids Avoid Wearing Hoodies?." In G. Yancy & J. Jones (eds.), *Pursuing Trayvon Martin: Historical Contexts and Contemporary Manifestations*, 129–40. Washington: Lexington Books.

Jewett, C. (1982). *Helping Children Cope with Separation and Loss*. Cambridge: Harvard Common Press.

Johnson, M. (2016). "Cooperation with Multiple Audiences." *Croatian Journal of Philosophy*, 16 (47): 203–27.

Johnson, M. (2017). "Seeking Speaker Meaning in the Archaeological Record." *Biological Theory*, 12 (4): 262–74.

Johnson, M. (2019). "Making Meaning Manifest." *Croatian Journal of Philosophy*, 19 (57): 497–520.

Johnson, M. (2019). "Must We Mean What We Wear?." *Aesthetics for Birds*, October 19. Accessed online.

Johnson, M. (2020). "Archaeology Excavates the Layers of Meaning We Leave Behind." *Aeon-Psyche*, November 4. Accessed online.

Johnson, M. (2021). "Adorning Intentions." *Journal of Aesthetics and Art Criticism*, 79 (4): 504–507.

Johnson, M. (Forthcoming). "What Bodies Mean." *The Philosophers' Magazine*.

Johnson, M. and Everett, C. (2021). "Embodied and Extended Numerical Cognition." In A. Killin and S. Allen-Hermanson (eds.), *Explorations in Archaeology and Philosophy*. New York: Synthese Library. Springer, 125–48.

Johnson Prum, A. (2010). *Nature: Hummingbirds - Magic in the Air* [Film].

Johnson Prum, A. (2016). *Super Hummingbirds*. [Film]. Nature.

Jones, A. and N. Ratterman (2009). "Mate Choice and Sexual Selection: What Have We Learned Since Darwin?" *PNAS*, 106 (1): 10001–8.

Joyce, C. (2010). "Study: Neanderthals Wore Jewelry and Makeup." NPR, January 12. Accessed online.

Judge, T. and D. Cable (2004). "The Effect of Physical Height on Workplace Success and Income: Preliminary Test of a Theoretical Model." *Journal of Applied Psychology*, 89 (3): 428–41.

Judge, T. A., Buho, J. E., Eret, A., Locke, E. A., (2005). "Core Self-Evaluations and Job and Life Satisfaction: The Role of Self-Concordance and Goal Attainment." *Journal of Applied Psychology*, 90 (2): 256–68.

Kant, I. (1790/1987). *Critique of Judgment*, trans. W. Pluhar. Indianapolis: Hackett.

Kaplan, D. (1989a). "Demonstratives: An Essay on the Semantics, Logic, Metaphysics, and Epistemology of Demonstratives and Other Indexicals." In J. Almog, J. Perry, and H. Wettstein (eds.), *Themes from Kaplan*, 481–564. Oxford: Oxford University Press.

Kaplan, D. (1989b). "Afterword." In J. Almog, J. Perry, and H. Wettstein (eds.), *Themes from Kaplan*, 565–614. Oxford: Oxford University Press.

Kaplan, I. (2019). "The Leopard-Print Midi Skirt Is the Summer Trend That Won't Die." *New York Times*, August 5. Accessed online.

Kecskes, I. (2014). *Intercultural Pragmatics*. Oxford: Oxford University Press.

Kecskes, I. (2016). "Intracultural Communication and Intercultural Communication: Are They Different?" *Lecture*, June 10. University of Split, Croatia. *7th International Conference on Intercultural Pragmatics and Communication*.

Khazan, O. (2019). "I Was Never Taught Where Humans Come From: Many American Students, Myself Included, Never Learn the Human Part of Evolution." *The Atlantic*, September 19. Accessed online.

Kiecolt-Glaser, J. K. et al. (1984). "Psychosocial Modifiers of Immunocompetence in Medical Students." *Psychosomatic Medicine*, 46: 7–14.

Kiecolt-Glaser, J. K. Glaser, R., Cacioppo, J. T., Malarkey, W. B. (1998). "Marital Stress: Immunologic, Neuroendoctine, and Autonomic Correlates." *Annals of the New York Academy of Sciences*, 840: 656–63.

Kingma, B. and W. van Marken Lichtenbelt (2015). "Energy Consumption in Buildings and Female Thermal Demand." *Nature Climate Change*, 5: 1054–6.

Knight, C., Power, C., Watts, I. (1995). "The Human Symbolic Revolution: A Darwinian Account." *Cambridge Archaeological Journal*, 5, 75–114.

Koda, H. and A. Bolton (2007). *Poiret*. New York: Metropolitan Museum of Art.

Koda, H., Murai, T., Tuuga, A., Goosseps, B., Nathan, S. K. S. S., et al. (2018). "Nasalization by *Nasalis larvatus*: Larger Noses Audiovisually Advertise Conspecifics in Proboscis Monkeys." *Science Advances*, 4 (2): 1–6.

Koppel, N. (2007). "Are Your Jeans Sagging? Go Directly to Jail." *New York Times*, August 30. Accessed online.

Kripke, S. (1977). "Speaker's Reference and Semantic Reference." *Midwest Studies in Philosophy*, 2 (1): 255–76.

Kripke, S. (1980). *Naming and Necessity*. Cambridge: Harvard University Press.

Kuhn, S. and M. Stiner, (2007). "Paleolithic Ornaments: Implications for Cognition, Demography and Identity." *Diogenes*, 54 (2), 40–8.

Lakoff, G. and M. Johnson. (1980/2003). *Metaphors We Live By: With a New Afterword*. Chicago: University of Chicago Press.

Lande, R. (1980). "Sexual Dimorphism, Sexual Selection, and Adaptation in Polygenic Characters." *Evolution*, 34 (2): 292–305.

Latour, B. (2004). "How to Talk About the Body? The Normative Dimensions of Science Studies." *Body & Society*, 10 (2–3): 205–29.

LeDoux, J. and R. Brown. (2017). "A Higher-Order Theory of Emotional Consciousness." *PNAS*, 114 (10) 2016–2025.

Leroi-Gourhan, A. (1968). *The Art of Prehistoric Man in Western Europe*. New York: Thames & Hudson.

Levine, P. A. (1997). *Waking the Tiger: Healing Trauma*. Berkely: North Atlantic Books.

Lewis-Williams, D. (2002). *The Mind in the Cave*. New York: Thames & Hudson.

Linton, E. L. (1883). *The Girl of the Period and Other Social Essays*. London: Richard Bentley and Son.

Lurie, A. (1983). *The Language of Clothes*. New York: Holt.

MacFarquhar, L. (2018). "Mind Expander: A Philosopher Asks Where We Begin and Where We End." *The New Yorker*, April 2: 62–73.

Maclean, D. (2019). "Americans Tend to Elect the Tallest Person for President— Here's How the 2020 Candidates Would Fare." *Independent*, November 18. Accessed online.

Malafouris, L. (2016). *How Things Shape the Mind: A Theory of Material Engagement*. Cambridge: MIT Press.

Mallet, J. and M. Joron. (1999). "Evolution of Diversity in Warning Color and Mimicry: Polymorphisms, Shifting Balance, and Speciation." *Annual Review of Ecology and Systematics*, 30: 201–33.

Martyris, N. (2016). "'Luncheon in Fur': The Surrealist Teacup that Stirred the Art World." *NPR*, February 9. Accessed online.

May, R. A. B. (2014). *Urban Nightlife: Entertaining Race, Class, and Culture, in Public Space*. New Brunswick: Rutgers University Press.

May, R. A. B. (2015). "Discrimination and Dress Codes in Urban Nightlife." *Contexts*, 14 (1): 38–43.

McNiel, P. (2018). *Pretty Gentlemen: Macaroni Men and the Eighteenth-Century Fashion World*. New Haven: Yale University Press.

McAfee, M. (2015). "Amber Rose's SlutWalk Sparks Debate." CNN, October 4. Accessed online.

McKinley, Jr., J. C. and R. Rojas (2016). "The Lives and Lies of a Professional Imposter." *New York Times*, February 4. Accessed online.

McMurray, R. G., Berry, M. J., Hardy, C. J., Sheps, D. S. (1988). "Physiologic and Psychologic Responses to a Low Dose of Naloxone Administered During Prolonged Running." *Annals of Sports Medicine*, 4 (1): 21–5.

McNeil, J. K., Le Blunc, E. M. Joynet, M. (1991). "The Effect of Exercise on Depressive Symptoms in the Moderately Depressed Elderly." *Psychol Aging*, 6: 487–8.

McNeil, P. (2018). *Pretty Gentlemen: Macaroni Men and the Eighteenth-Century Fashion World*. New Haven: Yale University Press.

McWhorter, J. (2003). *The Power of Babel: A Natural History of Language*. New York: Harper Perennial.

Mendes, K. (2015). *Slutwalk: Feminism, Activism, and Media*. London: Palgrave Macmillan.

Milam, E. L. (2011). *Looking for a Few Good Males: Female Choice in Evolutionary Biology*. Baltimore: Johns Hopkins University Press.

Minarik, J. (2021). "On the Adorning Arts." *Journal of Aesthetics and Art Criticism*, 79 (4): 493–498.

Ming, C. and G. Coakley (2017). "A Challenging Case of Fibromyalgia and Post Traumatic Stress Disorder." *Rheumatology Advances in Practice*, 1 (1 October).

Montague, R. (1974). Formal Philosophy Selected Papers of Richard Montague, ed. R. Thomason. New Haven: Yale University Press.

Morrigan, C. (2015). "Yes I Am a Slut." In A. Teekah (ed.), *This Is What a Feminist Slut Looks Like; Perspectives on the SlutWalk Movement*, 17–19. Ontanio: Demeter Press.

Mulkern, A. (2016). "Boots: Capitol Hill Cowboys." *Denver Post*, May 7. Accessed online.

Nagel, T. (1974). "What Is It Like to Be a Bat?" *Philosophical Review*, 83 (4): 435–50.

Neale, S. (1992). "Paul Grice and the Philosophy of Language." *Linguistics & Philosophy*, 15: 509–59.

Neale, S. (2007). "On Location." In M. O'Rourke and C. Washington (ed.), *Situating Semantics: Essays on the Philosophy of John Perry*, 251–393. Cambridge: MIT Press.

Nguyen, L. and M. Jeng (2021). "A Wall Street Dressing Down: Always. Be. Casual." *New York Times*, August 2. Accessed online.

Noori Farzan, A. (2019). "Five Years Ago, Obama Was Blasted for Wearing a Tan Suit." *Washington Post*, August 28. Accessed online.

Nussbaum, M. (2012). "Looking Good, Being Good." *Philosophical Interventions: Reviews 1986–2011*, 138–48. Oxford: Oxford University Press.

Paley, W. (1863/2009). "Natural Theology." In J. Perry, et al. (eds.), *Introduction to Philosophy: 5th Edition*, 46–51 Oxford: Oxford University Press.

Pager, T. (2018). "Can an Office Temperature Be 'Sexist'? Women, and Science, Say So." *New York Times*, August 28. Accessed online.

Pappas, N. (2008). "Fashion Seen As Something Imitative and Foreign." *British Journal of Aesthetics*, 48 (1), 1–19.

Pappas, N. (2015). *The Philosopher's New Clothes: The Theaetetus, the Academy, and Philosophy's Turn Against Fashion*. Oxfordshire: Routledge.

Pappas, N. (2017). "Anti-fashion: If Not Fashion, then What?." In G. Matteucci and S. Marino (eds.), *Philosophical Perspectives on Fashion*, 73–90. London: Bloomsbury.

Peirce, C. S. (1998). *The Essential Peirce. Selected Philosophical Writings Volume 2*. Bloomington: Indiana University Press.

Pennell, J. (2020). "7 Things You're About to See at Every Wedding: The Bridal Pantsuit is Here to Stay." *Today*, February 5. Accessed online.

Pennisi, E. (2019). "Evidence Mounts that Gut Bacteria Can Influence Mood, Prevent Depression." *Science*, February 4. Accessed online.

Pietrosky, P. (2018). *Conjoining Meanings: Semantics Without Truth Values*. Oxford University Press.

Pike, A. W. G., Hoffman, D. L., Garcia-Diez, M., Pettitt, P. B., Alcolea, J., et al. (2012). "U-Series Dating of Paleolithic Art in 11 Caves in Spain." *Science*, 336 (June).

Pinker, S. and P. Bloom (1992). "Natural Language and Natural Selection." In J. H. Barkow et al. (eds.), *The Adapted Mind: Evolutionary Psychology and the Generation of Culture*, 451–93. Oxford: Oxford University Press.

Pitts-Taylor, V. (2007). *Surgery Junkies: Wellness and Pathology is Cosmetic Culture*. New Brunswick: Rutgers University Press.

Plunkett, D. (2015). "Which Concepts Should We Use? Metalinguistic Negotiations and the Methodology of Philosophy." *Inquiry*, 58: 828–74.

Preucel, R. W. (2010). *Archaeological Semiotics*. Hoboken: Wiley-Blackwell.

Price, M. (2020). "Africans Carry Surprising Amount of Neanderthal DNA." *Science*, January 30. Accessed online.

Prinz, J. (2006). *Gut Reactions: A Perceptual Theory of Emotion*. Oxford: Oxford University Press.

Prinz, J. (2008). "Is Consciousness Embodied?." In P. Robbins and M. Aydede (eds.), *Cambridge Handbook of Situated Cognition*, 419–36. Cambridge: Cambridge University Press.

Prum, R. (2013). "Coevolutionary Aesthetics in Human and Biotic Artworlds." *Biological Philosophy*, 28: 811–32.

Prum, R. (2017). *The Evolution of Beauty: How Darwin's Forgotten Theory of Mate Choice Shapes the Animal World—and Us*. New York: Doubleday.

Pugsley v. Sellmeyer. (1923). 250 S.W.538 (Ark.) Clay Circuit Court, Western District.

Puts, D., S. J. C. Gaulin and K. Verdolini (2006). "Dominance and the Evolution of Sexual Dimorphism in Human Voice Pitch." *Evolution and Human Behavior*, 27 (4): 283–96.

Raby, P. (2001). *Alfred Russel Wallace: A Life*. Princeton: Princeton University Press.

Renfrew, C. (1994). "Towards a Cognitive Archaeology." In C. Renfrew and E. Zubrow (eds.), *The Ancient Mind: Elements of Cognitive Archaeology*, 3–12. Cambridge: Cambridge University Press.

Renfrew, C. and P. Bahn (2012). *Archaeology: Theory, Methods, and Practice*. Sixth Edition. New York: Thames & Hudson.

Richards, E. (2017). *Darwin and the Making of Sexual Selection*. Chicago: University of Chicago Press.

Riddell, F. (2019). "John Edmonstone the Former Slave who Taught Darwin." *Not What You Thought You Knew. With contributors Osborne, A. and Davidson, T. Sky History Podcast*, Nov. 26.

Rifkin, A. and S. Goodwin (2014). "Don't You Wish You'd Worn That? An Ironic Knit." *Tatler*. November. 70.

Ristau, C. A. (1990). "Aspects of the Cognitive Ethology of the Injury-Feigning Bird." In C. A. Ristau (ed.), *Cognitive Ethology: The Minds of Other Animals*, 91–126. East Sussex: Psychology Press.

Robins, G. (1993). "The Representation of Sexual Characteristics in Amarna Art." The Journal of the Society for the Study of Egyptian Antiquities, 23: 29–41.

Robins, G. (2008). *The Art of Ancient Egypt: Revised Edition*. Cambridge: Harvard University Press.

Roe, D. A. (1970). *Prehistory: An Introduction*. Oakland: University of California Press.

Rosenthal, D. (2006). *Consciousness and Mind*. Oxford: Oxford University Press.

Ruse, M. (2008). "The Origin of the Origin." In M. Ruse and R. Richards (eds.), *The Cambridge Companion to the "Origin of Species."*, 1–13. Cambridge: Cambridge University Press.

Ryle, G. (2000). *The Concept of Mind*. Chicago: University of Chicago Press.

Saint-Amand, P., Porter, C. A., Guynn, N. (1996). "The Secretive Body: Roland Barthes's Gay Erotics" *Yale French Studies* 90. Same Sex/Different Text? Gay and Lesbian Writing in French, 153–71.

Sánchez, E. (2017). *I Am Not Your Perfect Mexican Daughter*. New York: Ember.

Saul, J. (2002a). "What Is Said and Psychological Reality: Grice's Project and Relevance Theorist's Criticisms" *Linguistics and Philosophy*, 25 (3): 347–72.

Saul, J. (2002b). "Speaker Meaning, What is Said, and What is Implicated." *Nous*, 36 (2): 228–48.

Savage, D. (2007). "My Other Dog's a German Shepherd." *What I Learned from Television*. This American Life. Accessed online.

Scheel, D. (2016). "Signal Use by Octopuses in Agonistic Interactions." *Current Biology*, 26: 1–6.

Schiffer, S. (2016). "Philosophical and Jurisprudential Issues of Vagueness." In G. Kol and R. Poscher (eds.), *Vagueness and the Law: Philosophical and Legal Approaches*, Oxford University Press, Oxford; 23–48.

Scott-Phillips. (2015). *Speaking Our Minds: Why Human Communication is Different, and How Language Evolved to Make It Special*. London: Palgrave Macmillan.

Scutt, D., Manning, J. T., Whitehouse, G. H., Leinster, S. J., Massey, C. P. (1997). "The Relationship Between Breast Asymmetry, Breast Size, and the Occurrence of Breast Cancer." *British Journal of Radiology*, 70 (838): 1017–21.

Searle, J. (1969). *Speech Acts*. Cambridge: Cambridge University Press.

Senghas, A. (2004). "Children Creating Core Properties of Language: Evidence from an Emerging Sign Language in Nicaragua." *Science*, 305 (5691): 1779–82.

Shannon, C. (1971). *The Mathematical Theory of Communication*. Champaign: University of Illinois Press.

Siossian, E. (2018). "Satin Bowerbirds Fall Victim to Plastic Waste, Wildlife Experts Urge Mindfulness." ABC News (Australia), October 5. Accessed online.

Skyrms, B. (2010). *Signals: Evolution, Learning, and Information*. Oxford: Oxford University Press.

Smith, E. J. (2021). *Literary Slumming: Slang and Class in Nineteenth-Century France*. Washington: Lexington Books.

Sperber, D. and D. Wilson (1986). *Relevance: Communication & Cognition*. Hoboken: Wiley-Blackwell.

Sperber, D. and D. Wilson (2015). "Beyond Speaker's Meaning." *Croatian Journal of Philosophy*, 15 (44): 117–49.

Spranklen, A. (2020). "The Story Behind Princess Diana's Wedding Dress." *Tatler*, June 24. Accessed online.

Stafford, A. (2005). "Afterword: Clothes, Fashion, and System in the Writings of Roland Barthes: 'Something Out of Nothing.'" In Translated from French by A. Stafford. A. Stafford and M. Carter (eds.), *The Language of Fashion*. London: Bloomsbury Academic.

Stanley, J. and Z. Szabo (2000). "On Quantifier Domain Restriction." *Mind and Language*, 15 (2 & 3): 219–61.

Steele, V. (1984). "Review of *American Beauty: A Social History Through Two Centuries of the American Idea, Ideal, and Image of the Beautiful Woman* by Lois Banner." *Journal of Social History*, 18 (2): 300–2.

Steinhauer, J. (2019). "V.A. Officials, and the Nation, Battle an Unrelenting Tide of Veteran Suicides." *New York Times*, April 14. Accessed online.

Stiner, M. (2014). "Finding a Common Bandwidth: Causes of Convergence and Diversity in Paleolithic Beads." *Biological Theory*, 9 (1), 51–64.

Strathern, A. and M. Strathern (1971). *Self-Decoration in Mount Hagen*. London: Duckworth.

Sundstrom, B. (2017). "Why Male House Finches Can Be Different Colors." *Audubon Society BirdNote. Podcast*. Accessed online.

Svendsen, L. (2006). *Fashion: A Philosophy*. London: Reaktion Books.

Taylor, D. B. (2020). "For Black Men, Fear That Masks Will Invite Racial Profiling." *New York Times*, April 14. Accessed online.

Taylor, S. R. (2018). *The Body Is Not an Apology*. Oakland: Berrett-Koehler Publishers.

Tekiela, S. (2000). *Birds of New York: Field Guide*. New York: Adventure Publications.

Thomasson, A. (2017). "Metaphysical Disputes and Metalinguistic Negotiation." *Analytic Philosophy*, 58: 1–28.

Tinker v. Des Moines Independent Community School District. (1969). 393 U.S.503.

Tomasello, M. (2010). *Origins of Human Communication*. Cambridge: MIT Press.

Tracy, J. (2020). "Find Something Morally Sickening? Take a Ginger Pill." *Aeon*, February 21. Accessed online.

Traister, R. (2011). "Ladies, We Have a Problem." *The New York Times*, July 20. Accessed online.

Trebay, G. (2014). "Rewriting the Rules." *Men's Fashion Paris. International New York Times*. Friday, June 27. p. 9.

Trigger, B. (2006). *A History of Archaeological Thought*. Cambridge: Cambridge University Press.

Trivers, R. (2014). *The Folly of Fools: The Logic of Deceit and Self-Deception in Human Life*. New York: Basic Books.

Twomey, E., Yeager, J., Brown, J. L., Morales, V., Cummings, M., Summers, K. (2013). "Phenotypic and Genetic Divergence among Poison Frog Populations in a Mimetic Radiation." *PLOS One*, 8 (2): e55443.

Tye, M. (1994). "Sorites Paradoxes and the Semantics of Vagueness." *Philosophical Perspectives*, 8: 189–206.

Unnsteinsson, E. (2017). "A Gricean Theory of Malaprops." *Mind & Language*, 32 (4): 446–62.

Valenti, J. (2011). "SlutWalks and the Future of Feminism." *The Washington Post*, June 3. Accessed online.

Valles-Colomer, M., Falony, G., Durzi, Y., Tigchelaar, E. F., et al. (2019). "The Neuroactive Potential of the Human Gut Microbiota in Quality of Life and Depression." *Nature Microbiology*, 4: 623–32.

Van Der Kolk, B. (2015). *The Body Keeps the Score: Brian, Mind, and Body in the Healing of Trauma*. London: Penguin.

Van Overveld, T., de la Riva, M. J., Donázar, J. A. (2017). "Cosmetic Coloration in Egyptian Vultures: Mud Bathing as a Tool for Social Communication?." *Ecology*, 98 (8), 1–3.

Vellenga, R. E. (1970). "Behavior of the Male Satin Bower-Bird at the Bower." *Aust Bird Bander*, 8: 3–11.

Vellenga, R. E. (1980). "Distribution of Bowers of the Satin Bowerbird at Leura, NSW, with Notes on Parental Care, Development, and Independence of the Young." *Emu—Austral Ornithology*, 80 (3): 97–102.

Von Frisch, K. (1974). "Decoding the Language of the Bee." *Science*, 185 (4152): 663–8.

Wallace, F. (2018). "Selena Gomez admits that Yes, Her Met Gala Fake Tan Wasn't the Best It Could Have Been." Vogue Australia, May 9. Accessed online.

Wallace, J. (2019). *The Feather Thief: Beauty, Obsession, and the Natural History Heist of the Century*. London: Penguin.

Walsh, D. (2012). "Lawmaker Wearing Hoodie Removed from House Floor." *CNN*. Accessed online.

Weissner, P. (1983). "Style and Social Information in Kalahari San Projectile Points." *Society for American Archaeology*, 48 (2), 253–76.

Weymouth, M. (2014). "The Problem With SlutWalk's New Name." *Philadelphia Magazine*, September 23. Accessed online.

Wickler, W. (1968). *Mimicry in Plants and Animals*, trans. German by R. D. Martin. New York: McGraw-Hill.

Wilson, C. (2016). "Another Darwinian Aesthetics." *The Journal of Aesthetics and Art Criticism*, 75 (3): 237–52.

Wollan, M. (2013). "Fresno State Loves Its Bulldogs, but So Does a Gang." *New York Times*, November 7. Accessed online.

Wood, G. (2007). "Surreal Things: Making 'The Fantastic Real.'" In G. Wood (ed.), *Surreal Things*, 2–15. London: Victoria & Albert Publications.

Wragg Sykes, R. (2020). *Kindred: Neanderthal Life, Love, Death and Art*. London: Bloomsbury.

Wynn, T. and F. Coolidge (2004). "The Expert Neanderthal Mind." *Journal of Human Evolution*, 46 (4): 467–87.

Zahavi, A. and Zahavi, A. (1997). *The Handicap Principle: A Missing Piece of Darwin's Puzzle*. Oxford: Oxford University Press.

Zebrowitz, L. and J. Montepare (2005). "Appearance Does Matter." *Science*, 308 (5728): 1565–6.

Zebrowitz, L. A., Kikuchi, M., Fellovs, J. (2010). "Facial Resemblance to Emotions: Group Differences, Impression Effects, and Race Stereotypes." *Journal of Personality and Social Psychology*, 98 (2): 175–89.

Zilhão, J., Angelucci, D. E., Badal-García, E., d'Errico, F., Daniel, F., Dayet, L. et al. (2010). "Symbolic Use of Marine Shells and Mineral Pigments by Iberian Neandertals." *PNAS*, 103 (3): 1023–8.

Zilhão, J. (2014). "The Neanderthals: Evolution, Palaeoecology, and Extinction." In V. Cummings et al. (eds.), *The Oxford Handbook of the Archaeology and Anthropology of Hunter-Gatherers*, 191–213. Oxford University Press.

Index

Note: Page numbers followed by 'n' refer to notes.

Adler, Jonathan   184 n.1
adornment, meaning of   9, 10, 52, 104, 155
Adornment: What Self-Decoration Tell Us about Who We Are   12, 18
aesthetic intentions   19, 172
   requirement   172
aesthetics   79–80, 83–4, 160, 163, 164, 167, 168, 170, 173
   choices   159–82
   judgments   160, 164, 165, 169, 173
   preferences   79, 87, 91, 94–6
   restriction   50
Amarna period   113–15, 117
Amat, J. A.   135
American Beauty: A Social History   55
anaphora   46
Andersson, Malte   98, 135
androgynous   114, 116, 117
animal bodies   8, 71, 73–5, 77, 79, 83, 85, 131, 159
   Darwin on   71–86
animals   6, 27, 66, 71, 75, 81, 98, 131, 132, 136, 141, 161, 166, 168, 173, 176
   art   161–3
   behavior   6, 174–6
   minds   137, 175, 176
   traits   8, 71, 165, 168
   worlds   6–8, 10, 26, 28, 34, 57, 59, 63, 65, 67, 69
armband   5, 29–31, 44
art   176–80
The Artful Species   83, 162
artifacts   113, 118, 121, 124, 137, 172
artifice   55–6
artificial languages   46
artwork   104, 105, 115, 116
Ashwell, Lauren   49, 177
assumption   25, 40, 42, 43, 85–6, 156–8

Baldwin, James   105
Banner, Lois   55
Barnes, Elizabeth   11
Barthes, Roland   6, 20, 22–5, 118, 154
Barthes project   22, 24
Bates, Henry   75
Batesian mimicry   75
Baudelaire, Charles   56
beards   84, 86, 88–94, 97
beauty   10, 79, 88, 95, 159–82
   through choice   181–2
   and co-evolution   173–4
   and desire   167–8
Bezuidenhout, Anne   150, 151
Binford, Lewis   136
Biological Philosophy   160
biotic aesthetics   160
black armbands   7, 13, 21, 22, 29, 40, 51, 52
The Blind Watchmaker   82
bodies
   communication, homo sapiens   5–6
   depression   148–50
   interpretations   2–3
   in literature   143–4
   meaning in   1–13
   mind and   1–2, 145–6
   natural meaning and   34–5
   psychology of   146–50
   PTSD   146–8
   self and   149
   sexual dimorphism   168–9
   sexual selection   168–9
bowerbirds   134, 135, 139, 161–3
brain   10, 12, 140, 145–7
broken wing display   76–8
budget   26, 33, 35
bulldogs   30–1

Camp, Liz 53
Cantalamessa, Elizabeth 104
carotenoids 3, 133, 134
Carroll, Lewis 44
Cephalopod color change 131–2
Clinton, Bill 47
clouds 1, 32, 33, 35, 174
co-evolution 10, 11, 81, 99, 159, 160, 164, 165, 170, 171, 173–7
   and beauty 173–4
   natural and sexual selection, bodily adornment 99–102
   natural *vs.* non-natural meaning 174–6
   nightclub dress codes 103–4
   process of 99, 164, 174–7
   school dress codes 102
   Stonewall Riots 103
communications 3, 5, 7, 16, 20, 22, 24, 25, 31, 46, 48, 57, 69, 118, 128, 129, 138
consciousness 1–2, 43, 44, 142, 145
conscious process 104
conventions 7, 22, 30, 40, 49–53
cooperative principle 38–40
coronavirus pandemic 49
cosmetic coloration 133
crinolines 106–8
Cuomo, Andrew 101

Darwin, Charles 6, 71–3, 77–9, 83, 87–91, 141–4, 146, 161
   on animal bodies 71–86
   arguments 73, 77
   diversity in sexual preference 93–6
   Shakespeare and 144
   theory 81, 91, 181
Darwin and the Making of Sexual Selection 88
Davies, Stephen 12, 18, 19, 79, 83, 162, 163, 183 n.1
Dawkins, Richard 81–3
deceptive behaviors 76–7
decoration 19
Denkel, Arda 33
depression 140, 148
d'Errico, Francesco 124–6, 136, 138
Descartes, René 1
The Descent of Man 77, 87, 89, 161

Diamond, Jared 134, 161–2
Diener, Ed 32
Donnellan, Keith 37, 38, 44
dress
   and code model 22–5, 104, 154
   lying and deception 59–60

ecological theory 32
Emerson, Ralph Waldo 55
emotional states 9, 10, 59, 131, 139–44, 146
emotions 9, 10, 131, 132, 139–47, 149, 151
   expression 28, 139–41, 144, 145, 150
   man and animals, expression 140–3
England 21, 93, 94
The Evolution of Beauty 84, 160
explananda 20–1
expression 9, 127, 131, 139–58
   and adornment, Darwin's work 144–5
The Expression of the Emotions in Man and Animals 9, 139, 140–4, 146, 151

fashion 6, 7, 12, 17–19, 22, 23, 49, 50, 63–5, 154, 177
   adornment and 17–18
   slaves to fashion 49
Fashion: A Philosophy 43
fashion metaphor 88
The Fashion System 20, 23
Fawlty Towers 29, 44
feathers 4, 63, 78, 81, 87, 88, 98, 131, 133, 161, 181
female birds 81, 83, 87, 88, 137
female choice 80, 83–6, 89–91
Fernandez-Llario, Pedro 132
Fisher, Helen 146
Fisher, Ronald 80, 81, 84
flowers 12, 164, 173
Freeman, Jonathan 155
Friedman, John 85
Friedman, Vanessa 64
functions, adornment 16

genuine intention 43, 44, 49, 50
ghost in machine 2
Godfrey-Smith, Peter 67, 131, 132
Grice, Herbert Paul 5–7, 25, 26, 28, 31–3, 35–6, 38–40, 45–6, 175

Gricean view 37–56
  cooperative principle 38–40
  implicature 38–43

Handicap Principle 91–3
Hollander, Anne 18, 62, 63, 84, 97
Hoodfar, Homa 52
hoodies 152–5
House Finch Communicates 3–5
house finches 4, 5, 13, 71, 133
Hovers, Erella 111, 129
human beauty 84
human minds 173
human sexual selection 87–110, 168
  diversity in sexual preference 93–6
  female preference, traits 97–9
  Handicap Principle 91–3
  ladies and feathers 87–90
  reappropriation and metalinguistic negotiation 104–10
hummingbirds 34, 73, 164

I Am Not Your Perfect Mexican Daughter 42
imitative adornment, evolution 67
implicature 38–43
implicit reference 20
information 139–58
intentionalism 28, 37, 44, 48
intentions 7, 19, 26, 27, 30, 38, 40, 43–5, 47, 48, 59, 60, 123, 171–2, 175
invisible emotional expression 140, 145
janitor 59, 155–7

Jeffers, Chike 153
Jobs, Steve 64

Kant, Immanuel 169
Kecskés, István 150
Kiecolt-Glaser, Janice 146
Knight, Christopher 117, 119, 120
Kripke, Saul 38

Lande, Russell 83
Langton, Rae 49
language 2, 5, 7, 16, 22, 24, 43, 45, 46, 48, 50, 51, 118, 128, 150–3, 161, 170–1, 175
  affect and utterances 150–3
  philosophy of 2, 7, 16, 45, 49, 104, 151

The Language of Clothes 22
The Language of Fashion 23
Leroi-Gourhan, Andre 119
Linton, E. Lynn 106, 107
Lurie, Alison 22, 106

macaroni period 62, 63
McNeil, Peter 63, 85
McQueen, Alexander 179
Major, Hippias 167
Marbles, Elgin 63
Martin, Trayvon 152–4
"Mate Choice and Sexual Selection: What Have We Learned since Darwin?" 80
May, Reuben A. Buford 103
meaning. *See also* natural meaning
  of garments changes 154–5
  significance, prehistoric bodily adornment 137–8
  summary of types of 68–9
"Meaning" 25, 35
meaning-making 111
"mere" imitation 50
metalinguistic negotiation 9, 50, 104–10, 159
Miller, Barney 105
mimicry 67, 75–7
Minarik, Julia 180
minds 1–5, 10, 28, 135–7, 140, 145–8
  body and 145–6
The Minority Body: A Theory of Disability 11
misperception 139–58
mourning, black armbands 29
mud bathing practices 133
Müller, Fritz 75
Muslim Veil 51, 52

Nagel, Thomas 166
The Natural Background of Meaning 33
natural environment 5, 8, 71, 74, 98, 101, 134
natural meaning 6–8, 25–7, 31, 32, 34–6, 57, 60, 61, 67, 96–7, 113, 120, 145, 174
  and adornment 97
  bodies and 34–5
  false dichotomy 122–3
  imitation, ancient Egypt 113–17

imitation, middle ages   112–13
  imitation, prehistory   117–21
  imitation of   7, 8, 26–8, 57, 60–9, 96, 98, 113, 117, 120, 145
  instances of   31, 34
  interpreting prehistory   121
  nature of   35–6
  non-human animals   66–7
  vs. non-natural meaning   36
  nuanced account of   25–6
  Peircean interpretation, archaeology   121–2
  positive imitation of   61, 67
  red ochre   117–21
  requirements for   31–2
  truth and   32–3
natural selection   5, 8, 10, 71–5, 77–9, 98–100, 181
  and deception   74–5
  and mimicry   75–6
natural world   5, 8, 11, 67, 68, 73–5, 77, 78, 100, 126, 164, 181, 182
Neanderthal Adornment   135–7
Nefertiti, Queen   114, 115
Nicaraguan Sign Language   127, 128
nightclub dress codes   9, 103–4
Nixon, Cynthia   101
non-human animals   5, 66, 68, 71, 141, 160, 164, 166, 168, 169, 175, 182
nonlinguistic communication   7, 45–7
non-natural meaning   5–7, 25, 26, 28, 29, 31, 36, 38, 57, 61, 137, 174, 175
  in human adornment   123–7
Nussbaum, Martha   63

object of inquiry   15–16
ochre   111, 117–20, 122, 123, 125, 128, 129, 132, 136
offspring   68, 74, 81, 98, 119, 120, 161, 169
Oppenheim, Meret   179
ordinary language   7, 17, 57, 170, 171
The Origin of Species   4, 72–4, 77–9, 87, 93, 139

Paley, William   181
Pappas, Nick   50
peacock's feathers   78
Peirce, Charles   121
personal stylists   47–8

personal theory   11–12
Plato   50
pornography   169–71
predators   75–7, 81, 83, 99
Pretty Gentlemen   85
Prinz, Jesse   143
protected speech   21–2
Prum, Richard   6, 10, 79, 80, 84–5, 99, 101, 159–61, 163–5, 168, 170–3
  position, relevant questions   165–6
psychotherapy   146, 148–50
PTSD   146–8

Qafzeh caves   121–2, 129

Raby, Peter   100
races   74, 87, 88, 90, 93, 95, 96, 99, 151, 155–8
  adornment and perception of   155–8
radiating pleat effect   186 n.1
reappropriation   104–10
red ochre   85, 117–20, 123, 125, 128, 139
Relevance: Communication and Cognition   24, 25
Richards, Evelleen   63, 88
Ristau, Carolyn   76
rivals   66, 91, 162
Robins, Gay   114
runaway process   80
Ruse, Michael   74
Ryle, Gilbert   2

Sánchez, Erika   42
sartorial extravagance   85
satin bowerbirds   162
Saussure, Ferdinand de   23
Schiaparelli, Elsa   16
Schiaparelli's fashions   178
school dress codes   9, 102, 106
science of signs   23
semiotic theory   118
Senghas, Ann   127, 128
sexual dimorphism   168–9
sexual selection   8–10, 71, 72, 77–81, 83, 84, 87, 89, 96–7, 99, 159, 160, 165, 168–9, 181
  Darwin's theory of   79, 83, 91, 93, 181
  theory of   8, 71, 78–80, 93, 139, 161, 181
Shannon, Claude   25

shell beads 9, 111, 118, 123, 125–9, 139
Silverman, Kaja 85
SlutWalks 106–10
social communication 133
social perception 32
Socrates 167
speaker meaning 37–8
Sperber, Dan 24, 25, 184 n.1
Steele, Valerie 55
Stiner, Mary 127–9
Stonewall Riots 9, 103, 104, 153, 155
structuralism 6, 15–36, 117, 118
  vs. intentionalism 28
suppression 139–58
surrealism 179
Svendsen, Lars 17, 22, 24, 43, 44
symbolic meaning 9, 111, 123, 125, 126

tail feathers 78, 82
traits 71, 78, 80–1, 83, 96–100, 160, 165, 166
Tropical Nature, and Other Essays 79
true language 127

Ugly Sweater Party 41
unconscious selection 78–83
uniforms 48

Valenti, Jessica 108, 109
Valles-Colomer, Mireia 149
Van Der Kolk, Bessel 146, 147
Von Frisch, Karl 46, 129

Wallace, Alfred Russel 5, 74, 75, 79, 101
wallowing 132–5
Wilson, Deirdre 24, 25, 184 n.1
Wilson, Jeremy 58, 75
Wittgenstein 143
women's bodies 84, 86, 154
word meaning 37–8
Wragg-Sykes, Rebecca 136

youthful female beauty 84

Zilhão, João 136
Zimmerman, George 152–3
Zuckerberg, Mark 64

www.ingramcontent.com/pod-product-compliance
Lightning Source LLC
Chambersburg PA
CBHW062227300426
44115CB00012BA/2244